MODERN PLACER MINING

MODERN
PLACER MINING

M.J. Richardson

CONSOLIDATED
PLACER DREDGING CO.
Irvine, California, USA

Library of Congress
Catalog Card Number 92-90054

Consolidated Placer Dredging Co.
17951 Skypark Circle
Suite C
Irvine, California 92713, USA

Editing & Graphic Design by
Ron Bowman

Printed in South Korea

First Edition

DEDICATION

This book is dedicated to Norman Cleaveland, whose counsel and inspiration got me into the business in the first place. A creative mining engineer, his life has been absorbed in thinking-up new ideas to make tin, gold and diamond dredge mining more efficient. Inventing the Cleaveland Circular Jig was one of his many important innovations that has inspired others and brought him enduring admiration among placer mining engineers throughout the world.

PREFACE

Placer mining has taken on new dimensions in recent years as other minerals than gold, tin, platinum and diamond, found in unconsolidated sediments are mined with dredges. This has resulted in new technologies and types of dredges being developed and adapted.

The definition of "placers" was at one time confined to alluvial gold deposits. Refering to an early 1800's Spanish Dictionary, "Placer" was derived from the Spanish and placer mining in the European and African continents. After dredges began mining tin in Malaysia in 1922, the definition was expanded to include tin. Later, platinum and diamond was included as a placer.

It appears that the 1960's craze with manganese nodules at great depths in the oceans, started the widening of the definition to other minerals. Now placers are considered to be almost any mineral that is found in unconsolidated sediments, with the exception of sand and gravel. My feeling is that this loose definition of placers tends to confuse the layman including government agencies, and does nothing for the advancement of science in mining.

Another timely factor in placers that has led me to take up the task of compiling this book, is one of great import to the civilized world. That is the recent collapse of the Soviet communist regime and its apparent replacement at least in goals, by the free market system. As a result, some of the last remaining placer fields of significance may be opening up to outside development. Many obstacles remain to private mining companies being able to effectively enter those placer mineral areas but it is going to happen. Therefore, it is timely that MODERN PLACER MINING be documented.

Finally, there is a proliferation of "illegal" miners throughout developing countries mainly targeting gold and diamond. The consequence is a major wasting of resources, pollution, destruction of the environment and loss of life through careless mining practices. There is guidance available to the small miner and hopefully this book can help to serve that purpose through improved technology and practises in mining that will mitigate some of the damage.

Mort J. Richardson

Irvine, California
February 1992

BACKGROUND

The sources of data in this book are the cumulative result of experience emanating from CONSOLIDATED PLACER DREDGING CO., and its predecessor companies and founder, Frank Griffin (i.e., Griffin Co., Placer Management Ltd., Consolidated Purchasing & Design, Inc.). Leaders in placer mining beginning in the 1890's, were involved with CPD and its constant search for improved methods of evaluating and mining placer deposits of gold, platinum and diamond.

Frank Griffin graduated from Harvard University with a degree in Mechanical Engineering and was hired by the New Zealand dredge engineer, R.H. Postelthwaite, who brought the bucket ladder gold mining dredge system design to California from New Zealand in the 1880's. Griffin further developed that design into what became the "California-type" bucket ladder mining dredge (BL/M).

Griffin's participation in management, resulted in the formation of leading placer mining firms such as PLACER DEVELOPMENT LTD., YUBA CONSOLIDATED GOLD FIELDS, BULOLO GOLD DREDGING, PATO CONSOLIDATED GOLD DREDGING and NATOMAS, to mention the principal ones. Much of this was done while Griffin was a director in PLACER DEVELOPMENT of Canada, the original backer of many developments in that period in placer gold mining.

ACKNOWLEDGEMENTS

It is important that some of the individuals whose engineering and field experience contributed to the successful development of placer evaluation and mining techniques, be recognized. Few are still living, but this is a recognition of their important work. These include:

Frank Griffin, Charles A. Banks, L.A. Decoto, T.D. Harris, Col. R.E. Franklin, V.E. Bramming, J.D. Hoffman, James Gibson, George Crangle, E.I. Brown, Karl F. Hoffman, James S. Wroth, Norman Cleaveland, Patrick H. O'Neill, John Wells, Earl H. Malmstrom, but knowing that there were many others whose professional efforts contributed greatly to the improved knowledge of placer mining.

Mort J. Richardson

CONTENTS

DEDICATION   v

PREFACE   vii

BACKGROUND   ix

ACKNOWLEDGEMENTS   xi

CHAPTER I   1
EXPLORATION AND EVALUATION
Section 1 Prospect Determination   1
Initial Survey . Data Research . Aerial Surveys . Budget
Considerations . Evaluation Decisions.

Section 2 Evaluation Equipment   5
Gold Pan . Rocker . Sluice/Riffles . Portable Trommel.

Section 3 Drilling Equipment   8
Churn Drill . Hand Drills . Banka Drill Operation . Ward Drill
Operation . Engine Powered Drills . "Airplane" Drill .
Keystone Churn/Drill . Hammer Reverse Circulation Drill .
Background . Operation . Rotary RC Drill . Vibratory Drill.

Section 4 Bulk Sampling   19
Shafting . Washing All Material . Side Cuts . Cut Box . Method
of Sampling . Sample Treatment . Procedures . Pitting and
Trenching . Caisson Sampling.

Section 5 Equipment Selection   26
Scope of Project . Physical Conditions . Relationship to
Mining Methods . Arctic Regions . Equipment Comparison.

Section 6 Evaluation Procedures   29
Concentration . "Color" Estimation . Weighing and Prepara-
tions . Log Preparation . Abbreviations . Corrections &
Calculations . Drive Shoe Factor . Radford Factor . Final
Correction . Computer Program Evaluation(ELAS) . The
Option Supplement.

Section 7 Phases of Sampling   39
Theory of "Saturation" . Reconnaissance . Surface Samples .
Scout Drilling/Sampling . Selection of Equipment & Method
of Sampling. Terrain . Spacing of Holes . Indicated
Reserves . Equipment Selection. Grid Spacing & Saturation .
Development Drilling: Proven Reserves. Equipment
Selection.

Section 8 Physical Considerations of Prospects   48
Rivers and Streams . Ocean Regions . Tertiary Gravels .
Moraine Deposits . Desert Placers.

Section 9 Becker Drill Tests   52
Introduction . Test Procedures . Size Distribution . Percent
Distribution . Becker Factors . Conclusion . Linear
Regression . Core Rise . Ore Reserves . Cross-Sections .
Conclusions.
Figure 1-Rocker Schematic Drawing   60
Table A-Placer Sampling Equipment Comparison   61
Table B-Standard to Estimate (Au)-Colors   62-65
Table C-List of Abbreviations-On Drill Logs   66
Table D-Standard Values From Drill Hole Data   67
Sample Drill Logs   68-74
Sample Shaft Logs   75-77

Becker Test

Table 1-Tenor Comparisons   78
Table 2-Core Rise Measurements   78
Table 3-Core Volume Comparisons   79
Figures 1 to 4-Distribution of Gold by Wt./4 Size Ranges   80
Figures 5A & 6A-Linear Regression   81
Figures 5B & 6B-Linear Regression   82
Figures 5C & 6C-Linear Regression   83
Figures 7A & 7B-Distribution of Meaured Volumes   84
Figures 8- Line GK-E/W Cross-Section   85

CHAPTER II   87
FEASIBILITY ANALYSIS-
PLACER MINING DECISIONS

Section 1 Principal Elements For Analysis   87
Equipment Tradeoff Analysis . Production Cutoffs.

Section 2 Equipment Selection   92
Portable Wash Plants . Hopper Feed . Trommel
Screen/Classifier . Critical Questions . Economic
Analysis .
Table 1-Economic Analysis of a Gold Placer.   96

Section 3 BL/M Dredges-Large Scale Placer Mining   97
Dredge Mobilization
Figure 1-Placer Gold Mining-Equipment Tradeoff.   99
Figure 2-Cost of Operation Trends With Production Levels,   100
       Placer Dredge Mining.
Figure 3-Reserve Criteria vs. Tenor, Small Placer Gold   101
       Mines.
Figure 4-Bucket Ladder mining Dredges (BL/M) Production   102
       Experience.
Table 2-BL/M Dredge Optomization.   103
Figure 5-BL/M Dredges; Capital Cost vs. Production.   104
Figure 6-Mobilization Milestones.   106
Figure 7-Cash Drawdown Schedule.   107

ILLUSTRATIONS   107

Photo Figures

Figure P1.0 Ward Hand Drill   108
    P1.1 Land Operations
    P1.2 River Operations with Raft Tethered

Figure P2.0 Empire Drill and Crew, Brazil   109

Figure P3.0 "Airplane" Churn Drill, Drilling to 60 ft.,   110
    Philippines

Figure P4.0 "Keystone-type" Churn Drill, Sampling large   111
    gold placer, Grey River, New Zealand

Figure P5.0 Becker Hammer, Reverse-Circulation Drill,   112
    sampling to 110 ft, Grey River, New Zealand

Figure P6.0 Bulk Sampling by Shafting, San Antonio de Poto, 113
    17,000 ft Elevation, Peru

Figure P7.0 Bade Caisson Drill, 28" diameter, Bulk Sampling; 114
    Diamond Placer, Brazil

Figure P8.0 Bade Caisson Drill Components   115

Figure P9.0 Yost Klam Drill, Caisson Bulk Sampling;   116
    Truck configuration

Figure P10.0 BL/M Gold Placer Mining Dredges   117
    P10.1 14 ft^3, Rio Nechi, Antioquia, Colombia
    P10.2 9 ft^3, Nome, Alaska

Figure P12.0 BL/M Gold Placer Mining Dredges   119
     P12.1 10.5 ft³, Bulolo, New Guinea
     P12.2 11 ft³, Bolivia

Figure P13.0 BL/M Gold Placer Mining Dredges   120
     P13.1 14 ft³, Rio Nechi, Colombia
     P13.2 9 ft³, Pampa Blanca, Peru

Figure P14.0 BL/M Tin Placer Mining Dredges   121
     P14.1 24 ft³, Malaysia
     P14.2 14 ft³, Bolivia

Figure P15.0 Mining Dredges   122
     P15.1 BL/M 22 ft³, Tin, Indonesia
     P15.2 S/M, Sand & Gravel, 250 ft, Holland

Figure P16.0 Rutile Mining Dredges   123
     P16.1 CS/M with Processing Plant, Australia
     P16.2 CS/M 16"; Florida, USA

Figure P17.0 TH/M Dredges, Mining Sand & Gravel,   124
     English Channel

Figure P18.0 BL/M 20 ft³, Gold Placer Dredge; Mining   125
     200 ft. Below Surface, Yuba River, California, USA

Figure P19.0 BWS/M Dredge, Tin Placer Mining, Brazil   126

Figure P20.0 Dredge Mining   127
     P20.1 BWS/M Dredge, Rutile, Western Australia
     P20.2 CS/M Dredge, Sand & Gravel, Deep Digging

Figure P21.0 DS/M Dredge, Dead Sea Salts Mining, Jordan   128

Figure P22.0 S/M Dredges, Gold Placer Mining,   129
    Small Scale Operations

Figure P23.0 BL/M 14 ft^3 Dredges(2), Gold Placers,   130
    Large Scale Operation-Rio Nechi, Colombia

Figure P24.0 BL/M 12 ft^3 Dredge, Diamond Placer,   131
    MG, Brazil

Figure P25.0 BL/M Dredges, Placer Mining   132
    P25.1 20 ft^3, Gold Dredge, New Zealand
    P25.2 6 ft^3, Tin Dredge, Bolivia

Figure P26.0 BL/M Gold Dredges   133
    P26.1 Yuba 18 ft^3, 124 ft digging depth, California
    P26.2 6 ft^3, Dunkwa Gold Fields, Ghana

Figure P27.0 BL/M 12 ft^3, Tin Dredge, Burma   134

Figure P28.0 BL/M 9 ft^3, Tin Dredge "Sungei Pandan,"   135
    Malaysia

Figure P29.0 BL/M 24 ft^3, Tin Dredge, "Selangor #2,"   136
    Banka Island, Indonesia

Figure P30.0 BL/M 24 ft^3, Tin Dredge,   137
    "Perangsang #2," Malaysia

Figure P31.0 BL/M 22 ft^3, Tin Dredge,   138
    "Belitung #1," Indonesia

Figure P32.0 Dredge Mining "Pioneers," Newton   139
    Cleaveland (L), W.P. Hammon (R), on BL/M
    Gold Dredge, California, circa 1908

CHAPTER III   141
MINING SYSTEMS FOR PLACERS

Section 1 Categories of Placer Mining Equipment   141
Dredge Mining Equipment.

Section 2 BL/M Dredges for Placer Mining   143
BL/M Types . Design Comments of F.W. PAYNE & SON, by
J.A. Hewitt . Production Experience of BL/M Dredges .
Bucketline Speed. BL/M Dredge Subsystems . BL/M Dredge
Size Selection . Capital Cost of BL/M Dredges . Hull Design
& Construction . Digging System . Digging Ladder . Ladder
Rollers . Main Drive . Winch System . Main Hopper .
Trommel, Revolving Screen . Screen Design. Save-All
System . Circular Distributor.

Section 3 Mercury Amalgamation in Gold Placer Mining 158
Background . Chemical Composition . Characteristics . Method
of Use . Safe Systems . Recovery System W/Hg . Mineral Jig
Function. Jackpot Amalgam System . Handling of Hg. .
Summary.

Section 4 Other Bucket Dredges   165
Bucket Clamshell/Grab (BG/C/M) . Dragline . Backhoe
BB/M . Continuous Dragline Dredge.

Section 5 Bucketwheel Suction Dredge (BWS/M)   168

Section 6 Other Suction Dredges (CS/M, S/M, TH/M)   170
Cutter Suction (CS/M) . Ladder Pump . Plain Suction Dredge
(S/M). Trailing Suction Hopper Dredge (TH/M).

Section 7 Electronics for Dredge Mining   172
Automation "Fuzzy Logic" . Placer Mining Applications .
Control Systems.

(Section 7 Electronics For Dredge Mining Continued)

Figure 1-Elevation of 18 CuFt. Yuba Dredge No. 110.   174
Figure 2-Jackpot Amalgam System, Schematic Diagram.   175
Table 1-BL/M Dredge Production.   176-77
Table 2-BL/M Dredge Segregations.   178-79
Table 3-BL/M Dredge Size vs. Production.   180
Table 4-BL/M Dredge Hull Size Comparisons.   180
Table 5-Trommel vs. Dredge Size: BL/M.   180

CHAPTER IV   181
MINERAL JIGS

Introduction

Section 1 Milestones of Mineral Jig Development   182

Section 2 History of Mineral Jigs   183
Lode Mine Jigs . Power Jig . Moveable Sieve Jigs . Fixed
Sieve Jigs . Hand-Operated Jigs . Fixed Sieve Jigs.
Figure 1-Basket Sieves in 1500's.

Section 3 Laws of Jigging   185
Stroke Length of Plunger . Capacity of Jigs . Side-wall
effect . Efficiency Degrades with Distance . Overloading .
Size of Feed . Density of Mineral . Hindered Settling .
Jig Screens . Principles of Efficient Jigging.

Section 4 Placer Mining Applications   187
First Jig Installation on BL/M Dredge-1912 . Second Jig
Installation . Tin Dredge Recommendations . **Development
of Dredge Applications** . Natomas Tests . Pan American Jig
Development . Replacement of Riffles by Jigs . Yuba Jig
Development . **Tin Dredges in Malay States**; Early Jig
Installations . Circular Jig Development . Diamond Dredge
Mining Applications . Gold Dredging Applications . IHC
Holland Jig Development . MkII Cleaveland Circular Jig .
Mineral Jig Alternatives . Fallacy of Riffle Systems .
Arguments for Jigs.

Section 5 Selection of Mineral Jigs   198
Primary Factors . Example of Specifying Jigs . Flow Diagrams
of Placer Recovery Systems . Hutch Concentration . Jig Bed
Material . Jig Bed Screen.

Section 6 Operating Considerations   202
Adjusting Jigs . Exposure Time of Grains . Cycle Speed .
Length of Stroke . Hutch Water . Tuning of Jigs . Hutch Spigot.

Section 7 "Decelerating Flow of Material Over a Jig Bed   205
to Increase Production Capacity" by Norman Cleaveland
Figure 1-Sluicing gold and Jigging...De Re Metallica.   210
Table A-Mineral Jigs, Comparative Data, MK II Cleaveland
Circular Jigs.   211
U.S. Patent-Cleaveland Circular Jigs.   212-17

CHAPTER V   219
PLACER MINING OPERATIONS

Section 1 Mining Reserve Calculations   220

Section 2 Mining Course Design   221

Section 3 Personnel for Placer Mining   222
Project Failures . Litmus Test . Organization.

Section 4 Essentials of Operation   224

Section 5 Operational Examples   225
Gold Dredging in Colombia . Alaska Gold Dredging . Brazil
Tin Dredging . Brazil Diamond Dredging . New Guinea Gold
Dredging-Bulolo . Peru Gold Placer Dredging . Malaysia Tin
Dredging . Bolivia Placer Dredging . Indonesia Offshore Tin
Dredging . Holland Sand Dredging . Rutile Dredging . English
Channel S&G Dredging . California Gold Placer Dredging .
Dead Sea Salt Dredging . Small-Scale S/M Gold Dredging .
New Zealand Gold Dredging . Ghana Gold Dredging .
Conclusions.
Figure 1-Polygon Map of Large Gold Placer Reserves.   231
Figure 2-Organizational Chart, Placer Gold Dredging.   232

APPENDIX   233

CASE STUDIES   233

Appendix A   233
Bulolo Gold Placer Development Project, New Guinea
Early Beginning . Exploration of Bulolo . Exploration
Methods . Logistics Planning. Development Drilling .
Geological Theories . Dredging Systems.
Figure 1-Map of Goldfields, District of Morobe, Bulolo,   241
    New Guinea, 1934.
Figure 2-Ideal Cross-Section of a Placer Deposit Showing   242
    False Bottom.
Figure 3-Cross-Section of Bulolo River.   242
Figure 4-Mining Leases on the Bulolo , July 1931.   243
Figure 5-Dredge Courses, Bulolo and Bulowat.   245

Appendix B   246
Geomorphological Study, Gold Placer,
Nechi River, Colombia
Background . Introduction. Weathering and Transportation-
Geomorphic Applications to Economic Deposits . Residual
Deposits . Transported Deposits . Coastal Processes/Beach
Placers . Eolian (Wind) Processes . Glaciation. Mass-wasting
(Gravitational Processes) . Running Water (Fluvial
Processes) .

Regional Geomorphic Evolution .   255
Influence of Fluvial Processes . Tectonism and Terrace
Formation . Climatic Change and Terrace Formation.
Figure 1-Locational Map of Nechi River Area.   260
Figure 2-Nechi River & Reference Maps,   261
    Colombia & Medellin.
Figure 3-Location Map of Study Area & Tributaries to Nechi.   262
Figure 4-Nechi Channels and Floodplain.   263

Figure 5-Extension of Fig 4, Nechi Terraces.   264
Figure 6-Extension-Fig 4-5, Buried Terraces and Channel   265
    Gravels of Lower Nechi Floodplain.
Figure 7-(A) Diagram of Gravel & Coarse Alluvial Gold   266
    Deposition. (B) Typical Gold Concentration along
    inside of Migrating Meander Loop.
Figure 8-Hypothetical Paired Cut and Fill Terraces.   267
Figure 9-Downstream Convergence of Terrace and Modern   267
    Stream Gradients.
Table A-Auriferous Gravels of the Lower Nechi River   268
    Valley Near El Bagre.

Appendix C   269
Exploration of Au & Pt, Choco, Colombia
Background . Ward Drill . Evaluation of Boreholes by
4" Ward Hand Drills.
Figure 1-Choco Drilling Program Map.   273
Figure 2-Straight Thread-Type Drive Shoe.   274
Figure 3-Drill Log 1.   275
Figure 4-Drill Log 2.   276

Appendix D   277
BL/M Dredge Flow Diagrams/Processing Plants
Figure 1-Flow Diagram/Processing Plant, 20 ft^3 278
    BL/M Dredge, W/Cleaveland Circular Jigs
Figure 2-Flow Diagrams/Processing Plant, 14 ft^3 279
    BL/M Dredge, W/Cleaveland Circular Jigs
Figure 3-Portable Wash Plant 165 m^3/hr,   280
    Cleaveland Circular Jigs
Figure 4-Flow Diagram/Processing Plant, 18 ft^3 281
    BL/M Dredge, W/Yuba Jigs
Figure 5-Flow Diagram/Processing Plant, Yuba   282
    18 ft^3 Dredge, W/Cleaveland Circular Jigs
Figure 6-Flow Diagram/Processing Plant, 20 ft^3 283
    BL/M Dredge, W/Cleaveland Circular Jigs

Figure 7-Flow Diagram/Processing Plant, 14 ft^3 284
      BL/M Dredge, W/Cleaveland Circular Jigs
Figure 8-Flow Diagram/Processing Plant, 14 ft^3 285
      BL/M Dredge, W/Cleaveland Circular Jigs

Appendix E
HISTORY OF MINING ALLUVIAL GOLD   286
by Charles M. Romanowitz

BIBLIOGRAPHY   301

INDEX   307

CHAPTER I
EXPLORATION & EVALUATION

Section 1
Prospect Determination

Initial Survey

Before making positive plans on the selection of sampling methods, it is important that a complete physical survey be made of the prospect area. This inspection would normally be combined with a data search of prior activity of mining and prospecting that might provide a guide where the sampling locations should be and something about the subsurface conditions including depths to bedrock.

Most gold deposits have similar characteristics including the concentration of values occurring at or near bedrock. However, there are also variations and random occurrences of placer deposits that violate that principle. The Bulolo property in New Guinea assumptions based upon a "false" bedrock, which it was felt, resulted from the limitations of the drills. They were only able to bring in portable, hand-powered churn drills instead of engine-powered, due to the lack of roads (see Appendix Case Study A: Bulolo).

Once they introduced air cargo transportation and were able to bring in larger, engine-powered drills they penetrated the false bedrock and found the depths to be double what they had anticipated. Likewise, gold was contained in intermediate lenses at various depths above actual bedrock. A similar condition existed on the Yuba River in California where up to 22 dredges mined gold for 64 years (see Appendix-History/Romanowitz).

The consequence of the above variations in conditions of placer gold deposits is to place a heavy reliance upon experienced personnel to make initial evaluations. Background experience in a variety of conditions can be very useful when viewing a new area and deciding how best to approach it.

The occurrence of bedrock outcroppings, gravel and sand bars, terraces and paleo channels of meandering rivers and streams, give important indications to the experienced mining engineer. From such observances, he can evaluate the nature of the deposit and the best method to approach its initial sampling.

Data Research

Most areas of the world that contain significant amounts of placer gold deposits, have been identified and explored to some extent in years past. A review of mining libraries and companies involved in gold mining, may reveal information about the areas. While this type of information may be difficult to obtain at times, it can be very useful if available and may save a great deal in the exploration phase.

The libraries of government mining departments are usually a good starting point. The mining engineering societies in the USA, Britain, France, Spain and Portugal contain sources of data. Conferences and their proceedings on the subject of gold, particularly pre-1960, often contain placer mining histories.

The production records of gold dredges were usually published in local records, books, or periodicals that can be researched. Past production information that may show digging depths to bedrock, the nature of the material and the production of gold vs. yardage dredged can also be a useful aid in pre-planning. However, the best documents to find are copies of drill logs from early drilling, or at least summaries of their results.

Aerial Surveys

Satellite or aircraft-flown photography over areas that are remote and covered with vegetation, have been found to be useful by geologists and geomorphologists. In these cases they are able to detect geological anomalies using infrared, that can be correlated to base minerals. It has been useful for locating granite out-croppings in South America as prospects for cassiterite.

Other examples of useage has been in locating evidence of Kimberlite pipes for diamond in North America and Australia. Other recent examples have been reported of using magnetometers with low flying helicopters, to establish anomolies in evaluating diamond prospects.

To our knowledge aerial surveys have not been particularly useful for placer gold. The possibility does exist, however, for screening large areas for gravel deposits that may be hidden by jungle growth. Starting with known gold sources, this could be a means of planning an exploration drilling project.

Budget Considerations

The practical matter that will often govern the extent to which an evaluation may be carried out, is how much money there is to spend on the project. What is the overall budget for a potential mine, should it be proven? Is there a financial limit of the investor that must be considered prior to launching a program of sampling that may be out of proportion to his ultimate financial capability to follow through?

Exploration and development sampling can be expensive so these factors are important to determine in advance and will influence the approach to be followed in sampling. An under-financed project will often tend to restrict the sampling phase. It will proceed with minimum mining equipment in the hope that the promoter's descriptions, or other information obtained, is in fact correct.

This is one of the sources of grief in mining. The stories of "striking it rich," often eclipse the many projects that go fail. Even some of the best planned prospects have gone awry, but at least the chances of that happening are minimized by carrying out a proper evaluation. Therefore, considering the two ends of the spectrum, if a large investor has a property that has a potential of over 1.0 million ounces of gold, for instance, and if preliminary indications support that estimate, he should be prepared to mount a drilling effort that will both accelerate and saturate the property results. This might require an expenditure of over $1.0 million, including the drill.

At the other end, the investor(s) may see his capability of spending a grand total of $1.0 million to develop a small mine of an estimated 30,000 ounces. He should limit his sampling program to a budget of say $150,000 and be ready to stop if values do not come up to the minimum level for economic mining. Obviously, the two ends of the spectrum are not in proportion to the results but that only serves to dramatize the plight of the small miner whose capital limitations are usually finite and modest.

Evaluation Decisions
When the foregoing factors have been taken into account, the scope of the project should be coming into focus. This will determine to a large degree what will be the first step in physical sampling of the deposit. Part of the decision-making process should be influenced by how it is contemplated the deposit would be mined. The larger project of the above example involving an estimated 1.0 million ounces of gold, would normally involve multiple bucket ladder mining dredges(BL/M).

There could be a requirement for thawing permafrost as in Alaska, Northern Canada, China and Siberia. That would necessitate conducting preliminary studies as to thawing methods and their economics that might render the project uneconomic from the start. The tenor of the ground would influence the decisions.

Other possibilities of a large overburden, as in tertiary gravels, pose severe problems not the least of which is the sampling itself. Mining evaluation needs to be given serious consideration before proceeding too far with sampling. Out of this type of study, a rational decision can be made as to the best approach for selection of sampling equipment and procedures (Chapter II-Feasibility Analysis, will deal with this aspect in detail).

Section 2
Evaluation Equipment

Gold Pan
There has been considerable material published on the use of the gold pan, so I will not go into detail on the subject. For ease of portability and as a simulation of final recovery by large processing and recovery plants for gold, there seems to be no practical substitute for the pan.

All sampling systems described in this book come down to the job of panning the final concentrate in order to identify and count the "colors," or specs of gold, for logging and calculation. There are several varieties of gold pans available but it is important to select one that has smooth sides with no finish, lacquer or oil on its surface. Some professionals with international experience in placer sampling, tend to prefer the native "Batea," which is a carved, wooden bowl about 15 to 18 inches in diameter and 3 inches in depth at the center, with uniform slopes to a rounded bottom.

In the initial survey of a prospect, it is useful to take a pan to wash surface samples or samples taken from the bottom of pits and other diggings. This can at least establish the presence of gold on which to build confidence for the next phase of scout evaluation. A sieve pan with small 3/16-inch holes may be used as a preliminary screening device prior to panning.

Rocker

In the next phase of sampling after the initial survey where no more than surface samples are taken, the Rocker is the second level of useful tools (see Figure 1-Rocker Sketch). While the concept of a Rocker may be almost as old as the pan, it is another example of a method of final recovery simulation of large scale processing that is difficult to replace, particularly considering its low cost and portability with only manpower required to use it.

The Rocker is a <u>classifier</u> as well as <u>concentrator</u>, much as the mineral jig. All fine material, however, is captured for panning in the Rocker. Since more advanced alluvial gold mining operations use jigs exclusively without riffles or sluices, the Rocker becomes the ideal means of approximating that process and produces small quantities of black sands and gold (where present), for the final panning phase.

When operated by two men, a Rocker has a capacity of about three to five cubic yards in ten hours, using 100 to 800 gallons of water. The rocker may also be used in very small scale placer mining where the excavation is by shovel. But it is more appropriate when the volume of handling while using a high-speed drill such as the Becker, requires an active crew to keep up with the volume of samples produced.

Sluice/Riffles

In the process where bulk sampling is utilized, it is usually necessary to use a Sluice or referred to as the "Long Tom," to concentrate larger volumes of material. The sampled sand, gravel and boulders are flushed through in advance of running the material through a Rocker or panning for final concentration. The important consideration here is not to use too much water so as to lose the fine material but mainly to wash the larger rocks and gravels before discarding them.

This will reduce the amount of material either to a concentrate in the riffles or for further processing in the Rocker (see Illustrations-Figure P6.0). The size and length of the sluice will vary with the amount of material that is generated. The minimum size is about seven-foot-long and small enough to fit in the back of a pickup truck, or tied to a pack mule. An extension can easily be connected to it if more material is generated.

The sluice is not recommended by CPD for continuous mining operations because of its inefficiency in recovering fine gold. As a means of handling large amounts of material for sampling, it is another of the old, but still reliable tools for low cost, evaluation of placer gold deposits.

Portable Trommel
When conducting large scale, bulk sampling it sometimes becomes necessary to use a small trommel to classify and reduce the amount of material before final concentrations. For instance, when using a caisson clam drill a trommel may be needed in order to keep up with larger sample volume (see Figure P6.0).

The difficulty introduced with this mode of sampling, as with bulk sampling in general, is the inaccuracy of evaluation because of the chance that gold specs will become lodged in the equipment or simply washed out with the tailings. Sometimes it may become a pilot, small scale mining operation, using a jig along with the trommel.

In this case the evaluation takes on a different character and probability as far as blocking out a large area is concerned. A trommel for this purpose may be two to three feet in diameter, and the discharge through 1/2- to 1/4-inch diameter holes in the revolving screen. The undersized material may be concentrated further in a rocker to prepare the sample for panning.

Section 3
Drilling Equipment

It is not clear when the first drills were adapted to placer evaluation but the Banka in 1858 was certainly among the first. This must have been a great relief for the early miners who had been used to bulk sampling with pits and shafts, in the manner used for centuries. At the same time, however it gave great cause for suspicion as to its effectiveness.

The drill offered a more efficient means of reaching greater depths and in less time. It took the advent of the bucket ladder mining dredge to provide the real impetus for broader application of the drill to placers. Thus, sometime during the period of 1870-80, the Keystone drill, which had been used for water well and hard-rock minerals drilling, was applied to placer evaluation.

It wasn't until 1958 that a new concept of faster drilling was introduced that took a quantum jump beyond the various churn drills then on the market, with the Becker, Reverse-Circulation, Hammer drill. The actuation of the hammer provided the greater force needed and the extraction of the sample by compressed air or water decreased the time even further.

The new method was staunchly resisted by the old-timers and their disciples working in the placer business, and was therefore not until 1981 that the first drilling by a Becker drill into a placer that was subsequently dredge-mined, verified the reliability of the results. Those results are the subject of a report in Section 9.

Other equipment such as the Vibratory drill and the Rotary Reverse Circulation drill, are being used but there have been no proven results with mining to verify their accuracy, at this time. The essential consideration for drilling evaluation is to produce reliable results and it is therefore understandable that new methods are received cautiously. The following subsections will outline the features of each type of drilling system, how it operates and some of the considerations to be used in the selection of drills for a given property.

Churn Drill

The Churn Drill drives a casing into the ground as a means of obtaining an undisturbed sample. It uses a drive shoe ahead of the casing to force the material into the pipe and to provide a cutting edge. A bailer or suction pump, is dropped into the casing after driving it for some distance and extracts the sample that was pushed into the casing. There has been an evolution of churn drills that are presented as follows more or less in the chronology of their development.

Hand Drills

The first hand powered churn drill is recorded to have been used for alluvial tin sampling on Banka Island, east of Sumatra, Indonesia, circa 1858 and was named the "Banka" drill. In the early 1900's, the Empire drill was developed by an engineer with the **Pato** operations in Colombia and later was followed by the development of the Ward drill. All three drills operate on a similar principle, requiring a crew to work the auger (see Figures P1.0 & P2.0).

The total drill and pipe for 100-foot of depth can be transported by man or pack mule, broken down into components, and is therefore ideal for use in jungle or mountainous terrain. While the hand-powered churn drill is slow, it has been known to drill over 100 feet in a single-hole, and many thousands of cumulative feet of drilling have been accomplished using it in years past.

In swampy, jungle areas where a raft is required to cover areas that are under water, few other drills can replace this type of drill from an economical standpoint. It is limited by its slow speed and inability to cope with large boulders. It is ideal, however, for an initial scout drilling program in remote areas where cost is a factor.

Banka Drill Operation

The steps for operation of the Banka Drill, similar to Ward and Empire Drills, are as follows:

A "tongue" or auger drill is used to start the hole, attached to a section of drill rod. The drilling begins with two operators using the rod turner which is attached to the rod, turning the drill and boring down while a third operator watches to insure that the hole is plumb. When the drill rod has penetrated to its full extent, the tool is withdrawn and cleaned, then reinserted to clean the hole. The same process is repeated until ore is reached or until the hole starts to cave in.

The casing is inserted into the hole with a casing shoe attached. Taking care not to lose a casing, if the first hole is deeper than one length of casing, a double case wrench is attached to the top of the first casing to prevent it from falling into the hole. Another length of casing is then attached to the first length and tightened. This process is repeated with each successive length of casing until the bottom of the hole is reached.

When the casing has reached the bottom of the hole, a "platform socket" is attached to the top of the casing and a driller's platform is likewise connected. The platform is rotated several times to make sure that the parts are tightened and that the bottom of the hole has been reached.

The drill string is inserted into the casing and sufficient rods are added to reach the bottom of the hole. The "rod turner" is attached and two to four operators mount the drillers' platform and begin rotating the drill string. The same time the casing is rotated by means of several men standing on the ground, pushing on field cut timbers which are attached to the casing with a "casing grip." The casing shoe must be kept ahead of the drill string so that excess material is not recovered or drawn into the casing which would cause a "salting" of the sample.

The primary recovery tool is the sand bailer, or vacuum pump, which is fitted with a ball or flap valve, which allows the sample to enter, but does not permit it to fall back out. It may require three to four pumping procedures to remove the core, all of which is collected in a container for measurement. If hard material is encountered, a flat drill bit may be inserted in order to crush the material, which is then recovered by the bailer. This includes boulders, caliche or consolidated gravels.

When bedrock is encountered the casing is driven into the rock several inches and the hole cleanedout in order to retrieve the entire sample. In order to remove the drill string after reaching and completing the hole to bedrock, a pulling cap is attached to the top of the casing and a pulling stand is set up next to the casing.

A lever is inserted into the stand and attached to the pulling cap. Several operators are required to pull down on the lever while others agitate the casing to keep it free. As each section of casing is removed from the holes, it is uncoupled and the pulling cap is re-inserted into the next casing section. This process is repeated until all of the casing is withdrawn from the hole.

Ward Drill Operation

While the Ward Drill is almost identical to the Banka Drill, there are a few important distinctions (see Figure P1.0). The Banka Drill casing is rotated with the drill string auger, but the Ward Drill casing is stationary and is driven only by impact, using a man-powered spud arm. Using a winch, the spud arm is also used to drive and retrieve the drill string.

The framework of the drill can be made from timber or similar materials that may be locally available. The winch and walking beam or spud arm should be prefabricated prior to going into the field. The drill string, bits, casing and sand pump must be obtained from manufacturers. The Ward Drill is operated as follows.

Assemble the framework and winch system then attach the drill stem to a chisel bit and to the winch cable. A driving block (40 to 60 lbs.) is attached to the drill stem which is inserted into the casing. The top of the casing is fitted with a drive cap and the bottom with a drive shoe.

■ The walking beam is then raised and lowered causing a series of impacts to the casing, driving the casing into the ground. The casing needs to be monitored as it is driven to insure that it is plumb. As the hole is driven downward more casing is added, as necessary. The chisel bit loosens the material, which is recovered with a sand bailer and collected for volume measurement and final evaluation.

■ After completing a hole, the casing can be pulled by threading a pulling head on the top of the casing and replacing the chisel bit with a pulling hammer. Another method is to pull the casing from the bottom using a pulling jar and a casing spear, which grips the inside of the casing near the bottom. The advantage of the latter is that it does not place as much strain on the casing joints.

■ The Ward drill normally uses a 4-inch to 4-1/2-inch casing, which produces a small sample acceptable for preliminary investigation work. The Ward requires more labor than the Banka Drill because of the need for operating the walking beam, which produces the hammering action of the drill string and blocks. A crew of 12 to 15 men may be necessary to carry out an exploration program with the Ward Drill. On average, in tropical ground, 10 feet of hole-per-day can be expected from the Ward drill vs. 20 feet with the Banka.

Engine Powered Drills

The engine powered Churn drill employs a gasoline or diesel engine that turns an eccentric wheel to raise the block and drop it on the pipe. The Keystone drill, which is engine powered, was developed in Pennsylvania in the mid-1800's for water wells. It was used for testing hard rock minerals in the east and mid-west before being applied to placers in California, perhaps in the 1870's or 1880's.

This was followed by larger drills of the same type by Bucyrus-Erie, Koehring, and Hillman. While more cumbersome than the hand drills, the engine powered drills complimented and gave impetus to the development of the bucket ladder mining dredge by permitting more rapid evaluation of deeper and harder ground.

"Airplane" Drill

The smallest of the engine-powered churn drills is the "Airplane" Drill, so named because of its portability on small aircraft (see Figure P3.1-2). It usually has a 10hp gasoline engine to power the eccentric for lifting and dropping the weight to drive the casing into the ground. The drill structure can be easily disassembled to make it more portable and is useful for remote regions because of that. It can be carried by light truck, mule, raft or small aircraft as mentioned above.

It also can be moved short distances such as from one hole site to the next, by winching it over the ground. The average weight of the assembled equipment is 1800 lbs. The heaviest single piece weighs about 150 lbs. Drilling tools may add up to 500 lbs. The longest section of the drill is the derrick, which is 13 feet, but can be ordered in two pieces of 6-1/2 feet each.

The "Airplane" drill is designed to use either a 4" or 5" I.D. casing. Recovery of samples is accomplished with a sand pump. A crew of four to six men can do exploratory evaluation with this drill. The "Airplane" drill will work well in rocky ground to 50-foot depth. Greater depths can be achieved but progress becomes slow and may be difficult in tight formations. In average ground, 25 feet of hole-per-day, may be expected.

Keystone/Churn Drill

The Keystone-type (generally known as a churn drill), is considered the "work horse" of placer gold evaluation and has been used since at least the 1880's for placer evaluation (see Figure P4.1-3). It normally uses a 28hp engine and up to 40 hp on larger ones such as the Bucyrus-

Erie. There have been more properties tested by the churn drill than any other and therefore the validation of its results makes it difficult to introduce newer types of equipment. Many successful gold dredging operations were preceded by drilling with the Churn drill.

The Churn drill can use four-inch, five-inch or six-inch I.D. casings. Where there has been a large property, this type of drill has normally been used. The drill may be mounted on skids, truck, tractor treads or mounted on a floating barge to drill over water. This type of drill is produced by only a few manufacturers, but since the patents have long ago run out, any shop with the drawings may produce one.

The larger, engine-powered Churn drill operates on the same principle as the Ward drill. A casing is driven, the drill string loosens the material and the sample is retrieved with a sand bailer or vacuum pump. Hillman, Koehring-Speedstar, Bucyrus-Erie and Loomis are some of the designs that have been produced. However, the success with these drills depends upon the experience of the personnel in their use. The drilling crew may handle the functions, but there must be close coordination with a knowledgeable placer mining engineer on the project to produce reliable results.

One of the advantages of the powered Churn drill is the use of a smaller crew; one drill being operated by two men. Also, it can drill more easily to greater depths. In hard ground it may penetrate 20-25 feet-per-day, but in soft material perhaps twice as fast and deeper. The larger diameter sample taken is another advantage over the small hand drills.

Hammer, Reverse-Circulation Drill

Background

The Hammer, Reverse-circulation (RC) drill was first used by the Becker Drill Co., Canada, in 1958 (see Figures P5.1-.3). While it had slow acceptance by the placer gold industry, it was successfully used in alluvial tin in the 1960's. The latter application was conducted offshore Puhket, Thailand, where several thousand holes were drilled before going ahead to build a large tin mining dredge. The tin recovery by a second, replacement dredge validated the Becker drill results within 90% R/E (i.e. Recovery/Drill Results).

In the case of placer gold, tests were run in the 1960's using the Becker drill which included extensive drilling during the winters from ice floes offshore Nome, Alaska. The area was finally dredged using the "BIMA," a 30 ft^3 BL/M dredge in 1989-90, essentially validating the tenor of gold established by the above drilling.

The first instance of using the Becker drill on gold where dredging was accomplished afterwards to verify the results, was in 1981 in New Zealand, South Island, on the Taramakau River with the "Kanieri" dredge. Considerable drilling was accomplished alongside drill holes made in earlier years by churn drills. Verification was accomplished with the Becker drilling at the same time with two churn drills.

Statistical evaluations were made of the results of the two types of drills along with the dredging results. An adjusted "Drive Shoe Factor" was developed for the Becker drill, in that particular formation. Once applied the results of the Becker drilling correlated with a high degree of accuracy to earlier and current churn drill results (see Section 9 Report). It is understandable that the RC Hammer drill can cause a great deal of concern to those who have spent their lives using churn drills and verifying the accuracy of their results with the reliable method; which is to "mine it."

The use of high pressure air or water to flush out the sample, surging it through a cyclone and into the sample bucket, raises the question of whether some of the fine specs of gold may be "lost." It doesn't require a lot of calculations to determine the relative influence of a small spec of gold on the results from a drill that has an inside diameter of only 3-1/8 inches. Larger casings are used with a model of Becker drills with 4-1/8 inches I.D., but require longer periods for the same amount of penetration. Tests have shown, however, that both small spec's and large nuggets, are in fact not lost in Becker drilling.

Another problem that must be accomodated when using the Becker drill, is the speed of drilling which produces the samples at such a rate as to require an active sampling crew to keep up. Since it is important at all times to avoid possible "salting" of the prospect, it is a standing policy at CPD to avoid bagging samples and to conduct the reduction to gold specs and black sand on the site as the drilling progresses. Weighing and logging calculations are usually made after the drilling is completed that same day.

Operation
The Becker drill or comparable type of equipment is replacing the churn drill in many areas. Its weight of about 23 tons for the drill and truck, as well as requiring a water tank truck auxiliary alongside, makes it difficult to take into swampy or mountainous areas unless means can be found to prepare a suitable access. Some of the drills are mounted on tracks, as well as sea-going barges or jackup platforms.

The major advantage of the Becker Drill is its speed of penetration which has been found to be five to ten times greater than the churn drills. With the time saved the cost difference can be compensated for rapidly in terms of labor cost, including the ability to complete drilling and proceed with mining sooner. Even in the case of small, scout drilling projects if a rental drill is available, it can be cost effective.

There has been an active market in the U.S. for Becker drill rentals. In remote areas outside the U.S., however, the purchase of a used or new drill may be the only recourse to having one. There are also other manufacturers of RC Hammer drills. A crew of two men can normally operate a Becker drill but they must be well trained in its use. Since it is hydraulic and diesel powered, at least one member of the crew must be able to handle on-site repairs of the system. An additional two-man crew must be used for sampling or a total of four vs. two for a churn drill.

The pipe is double-walled and strong but costly, so an on-hand supply of extra pipe is necessary while drilling. The integral cast pipe has proven to be more durable and is recommended over the welded units. The Becker drill is a well-packaged system and while operating can be efficiently utilized. On one project when drilling in clean gravels to bedrock of 80-100 feet, as much as 300 feet of drilling was accomplished in a single-day.

Another advantage of the RC drill is the fact that a rotary drill bit can be inserted into the pipe for drilling through large boulders prior to hammering. It is effective to a depth of 140 feet and has been used to 340 feet. Depending upon the formation, an auxiliary compressor may have to be used when drilling below 130 feet.

In making calculations from a RC drill sample, it should be recognized that because the sample is continuously extracted using air or water under pressure, a "core rise" is not applicable as with the churn drill. Therefore, the volume of the sample must be measured in a suitable container to obtain volumes for grade calculations. Likewise, since the factors developed and verified with a dredge have only occurred in one location and operation, any extensive drilling should be checked with a churn drill prior to a final decision on a large investment. In other words, the "drive shoe factor" may vary from one formation to another, and thus needs re-calculation.

Rotary, Reverse-Circulation Drill

While there has been a great deal of controversy on the acceptability of the Hammer drill for placer evaluation, the use of the Rotary drill is very new to the scene and has had no verification as to its accuracy in recovery of placer gold samples. Recent use of it has been for drilling of deep, tertiary gravel deposits in Northern California, to 800 feet, through hard materials such as basalt and andecite.

The benefit of the drill is that it can auger through very hard materials in a reasonablly short period of time and test for bedrock as well as for the presence of gold. The use of a rotary drill is generally considered inadequate for placer sampling by experienced placer mining engineers. Drilling is done outside of a casing and therefore has no control over the amount of sample that is drawn in from outside of the hole. The exception is when drilling in permafrost where the core is brought up as a solid unit and thawed.

The rotary drill can therefore be considered useful for hard and deep penetrations but only to establish the presence of gold along with some assessment of the formation. However, it cannot be relied upon in its present configurations for an accurate sample. By employing a combination of casing with a rotary bit, we see where this type of drill could become useful in expediting the evaluation of bouldery and deep ground properties. Past testing by CPD using rotary drills in normal river gravels, proved to be unsuccessful due to high wear on the drill bits on rounded gravels.

Vibratory Drill

There were two types of vibratory drills introduced in the 1960's that were used both offshore and inland, in difficult conditions such as permafrost in the northern climes: AMdril System and the Vibrocorer. The AMdril system provides a disturbed, but quantitative sample using Low frequency vibrations; the Vibracorer provides undisturbed samples in selective material using higher frequency vibrations.

We are aware that vibratory drills have been used in prospecting for alluvial gold in Alaska and are told of problems with penetration when encountering boulders and other hard material; that the cost of operation is high, and that maintenance is a problem.   To our knowledge, there is no proven gold dredging operation that has verified the results of vibratory drilling, and therefore we present this as a possible use for future verification.

Section 4
Bulk Sampling

The oldest method of bulk sampling, going back to the earliest recorded mining, is by hand digging of vertical shafts large enough to accomodate a laborer (see FigureP6.1-.2). Where low-cost labor is prevalent, this can often be the most expeditious and economical method of sampling when the ground is not too deep. When depths exceed 30 feet, this method is not only impractical but may be dangerous to the workers. Shafting is helpful to:

■ Evaluate the size distribution of boulders and gravel, the presence of clay and the general appearance of the soil for planning of excavation methods.

■ Achieve a larger sample in a given area.

■ Check on drilling results (not necessarily a valid check since shaft sampling has its own inaccuracies).

Shafts can be dug by three or four men efficiently, to depths of 30 feet. The rate of digging can vary from three to six feet-per-day for a three-foot by three-foot shaft. Depending upon the compactness of the soil, the shaft should normally be cribbed with timbers or metal caissons, which will prevent caving and sluffing that can affect the accuracy of the samples below. Shafting is usually limited to elevated terraces where a water table will not be encountered or where a pump can be used to maintain sufficient dryness to work.

Shafting

New shafts may be sampled by any one of three methods, as follows:

- Washing up the entire amount of material excavated.

- Side cuts of measured width and depth down one or more sides of the shaft.

- Continuous sampling from the bottom of the shaft with a Cut Box.

Washing All Material

This will only be used where an abundance of water is available or as a check against the second or third methods on selected shafts. In this method the material will be excavated and washed in three-foot vertical sections and the volume calculated from measurements made of the shaft itself. A uniform shaft size from top to bottom should be maintained. When bedrock is reached, the depth of the sample sections should be reduced to one foot instead of three feet, as this is usually the most important area.

Side cuts

This method should be used in sampling shafts, which are first excavated by contract or unsupervised labor, or in re-sampling existing shafts. In this method a cut one foot deep and one-foot-wide, should be taken from surface to bedrock on one or more sides of the shaft and in three-foot sections. Each section should be washed separately. Cuts should be continuous and not offset horizontally, if it can be avoided.

Where drifts or old workings are encountered, it may be necessary to move to another wall of the shaft or to continue the cut on the face of the drift itself. Where large boulders are encountered in a cut, they should be removed or displaced, but all fine material scraped off of them within the limits of the cut.

Their volume should be included in the bank measurement of the sample but will not be included in the box measurement.   Large boulders that are excavated, if well brushed off, need not be taken from the shaft, but should be estimated for screen analysis percentages.

The greatest care must be taken to catch all fine material from the sample cut, and overcutting in depth or width must be avoided. For this reason, it is preferable to leave projecting boulders, which are partly in and partly out of the cut, in place. If they are dislodged, there is too much chance of an extra amount of fine material getting into the sample and over-valuation resulting.

The side of the shaft to be sampled should first be thoroughly cleaned, from top to bottom, and all loose material removed. This will avoid "salting" errors which might be caused by re-concentration at exposed points. The bottom of the shaft should then be thoroughly cleaned out and a canvas laid down of sufficient size to catch all fines from each section cut.

The engineer in charge should personally measure the width and depth of sample cuts after excavation and apply a corrective factor to the sample if overcutting or undercutting is found. Such corrections should be noted on the log sheet in all cases. All values are calculated on the overall cut or bank measurement, and not on the box measurements.

Cut Box

The Cut Box method is particularly useful in sampling loose or fine gravel deposits while a shaft is being sunk. It is, however, difficult to use in ground carrying a large percentage of oversize boulders six inches or more in diameter. This method should always be used in ground which will not stand up well in the shaft walls or in wet ground, as it avoids the possibility of oversize samples.

Method of Sampling

The general procedure is to set a steel "cut box" on the surface of the ground (and later on the bottom of the shaft) and excavate the material inside it, driving the box down as material is taken out. The "cut box" is usually made of about eight- or ten-gauge steel, either square in section with outside measurements of one-foot by one-foot, or circular-section with equivalent area.

The box is open at both ends with sharp, smooth edges at the bottom and a one inch steel flange around the outside at the top. The overall height under the flange is one-foot. When the box has been driven down to its full depth of one-foot and all material inside excavated and placed in sample cans, a piece of steel plate about eight-inches-square and 3/16" thick is placed flat on the bottom of the excavation inside the box.

The box is then removed, the remaining area of the shaft excavated, and the material discarded. When the small steel plate is reached in excavating, the cut box is again put in place and the procedure repeated. In some cases it is not necessary to remove the "cut box" from position while excavating the shaft around it, but the bottom plate should always be used to ensure the removal of any outside material spilled into the box.

Sample Treatment

Shaft samples should in general be processed in a Rocker and the concentrate panned although when the entire shaft content is washed, a long-tom or sluice can be used to advantage. The Rocker oversize is discarded, after inspecting for nuggets, but the Rocker tailings should be processed a second time. Steel Rockers are preferable as they are easily washed out.

Coco-matting, corduroy, or blanket material, overlain with expanded metal, are all adequate gold savers for the bottom portion of the Rocker, but are hard to clean. Steel or wood riffles are efficient and easy to clean but produce an unduly large amount of concentrate for panning. It is largely a matter of choice or experience of the Rocker operator.

Procedures

The panner should not pan down to a clean concentrate, but only far enough to count his colors and estimate the weight of gold. When down to a black sand concentrate, all overpannings should be caught for measurement and repanning. It is often advisable to throw all overpannings back into the Rocker to be re-run with the Rocker tailings, depending largely on the character of the gold and the skill of the panner.

The concentrates from each field panning should be placed in separate tins or porcelain ointment jars, with an identification slip, for final cleaning and weighing at the camp. Mercury should be used to recover the gold and the heavy residue examined for possible traces of other metals. Amalgam may be either restored or reduced with dilute nitric acid and thoroughly washed. (Special care should be taken in retorting/vaporizing the Mercury-see Chapter III-Section 3). In either case, the gold should be heated and annealed, then examined for impurities before weighing.

All weighing and calculations should be made in milligrams after an assay for "fineness" has been secured on representative samples. Check assays should be made occasionally or at any time a change is noticed in the character or color of the gold. All samples should be retained until the examination is completed and final estimates made.

Before gold samples are finally discarded they should be grouped or combined by areas and a careful screen analysis made of each combined sample, using standard laboratory screens from 10 to 200 mesh. Such a screen analysis is a useful check against future recoveries similarly analysed.

Pitting and Trenching

The "modern" method of bulk sampling is accomplished by digging a pit or trench using backhoe or dozer. As with shafting, if low-cost labor is available, manpower may be preferable.

Normally, a backhoe is the best method of pitting or trenching since its action can make a uniform cut on both sides and the bottom. The difficulty with this method, as with shafting, is the accuracy of the sample and the considerable amount of material that must be processed. The more accurate procedure without the added cost of equipment, capable of processing the excavation, is to cut a vertical channel as described in Shaft sampling above.

The trenching method is often applicable to small streams with shallow ground 10-15 feet, and a single trench may reveal a great deal about the deposit. However, it will probably be necessary to process all of the excavated material since it may not lend itself to channel sampling. Using spacing between trenches of 500 to 1,000 feet may be sufficient to determine the extent and viability of small streams in confined areas, such as a narrow valley with shallow depths to bedrock.

This process will simulate small scale mining, and the use of a pilot washing plant with such a method may be appropriate. A great deal of work was done in earlier days of gold dredging, to determine the accuracy of bulk vs. drill sampling. The variations in results with some of the tests indicate that the methods used may contain errors, but on the whole, one seems to check out with the other.

Therefore, the decision of which one to use is more dependent upon depth of ground, availability of labor, equipment, cost of sampling, such as when using a washing plant for the processing, budget, and urgency of decisions. The presence of groundwater will eliminate most bulk sampling if water is encountered in large volume before reaching bedrock.

Caisson Sampling

The use of caisson digging equipment has produced a method of bulk sampling that overcomes the limitations mentioned above, such as water table and depth. Such equipment as the Bade Drill, made in Germany, or similar equipment in Canada, where a steel pipe varying in I.D. from 0.5-1.0 meter, is driven down by rotational action combined with hydraulic pressure, and the weight of the pipe. The sample is extracted by a clamshell bucket that fits the inside contour of the pipe (see Figures P7.0 and P8.1.2).

Caisson drilling provides a casing to maintain a uniform volume and preventing sluffing or erosion by water. The larger volume of material usually requires a small washing plant to process the sample. A long sluice can be used but may hold up the progress of the drill.

A mobile crane is required to handle the large pipe and that alone presents a problem of transport, access to remote areas and handling. Since the total cost of the system may be as high as $1.0 million, it is evident that the prospect must be of large proportion to justify using it. The caisson system has been used in deep tertiary gravels of Northern California and Peru for gold, Bolivia for tin, and in Brazil for diamond. The hardness of the subsoil, however, can cause delays. The "Yost Klam Drill" is perhaps the least expensive, practical caisson equipment that has been used to any great degree for placer gold, both in the U.S. and South America. However, its availablity is limited (see Figure P9.0).

Section 5
Equipment Selection

Scope of Project
The choice of which method and equipment to be used for sampling, whether by bulk excavation or drilling, can usually be placed into perspective from the following factors:

- By the area of the property to be evaluated;
- By the depth of ground;
- By the budget of the investor.

(CHAPTER II-FEASIBILITY ANALYSIS, develops this decision process).

Whether a large property or small, no one wants to spend more money at the speculative stage of a prospect than he has to. Once it has been established that there are potential quantitites of gold that could make a profitable mine, a higher level of confidence can be established that may justify more expensive and faster methods of sampling. These aspects are discussed in more detail in the following paragraphs.

Physical Conditions
The method of sampling and hence the equipment, is influenced by both the surface and subsurface conditions of the prospect. Encountering boulders and stiff clay can be a problem regardless of the climatic conditions, from the arctic to the tropics. Allowing for those conditions can be critical.

Relationship to Mining Methods

Some prospects have been grossly undervalued where the clay was not properly accounted for or conversely, "overvalued," which can be more dangerous from a financial standpoint. In all cases, it is important to have an experienced placer mining engineer involved in the equipment and method selection as well as with sampling.

In addition to anticipating obstacles, he needs to relate the sampling to the potential means of mining. Large scale placer mining by most standards, will have to involve dredging and the recovery system should be visualized during the sampling effort.

The ability of a dredge of a given size to mine the deposit is a matter of high importance. The treatment plant to recover gold from various difficult conditions whether in heavy clay or slimes, should be addressed so that pertinent, vital information will be recorded in the logs.

The accessibility of the prospect has an influence on the selection of the equipment. In remote regions, the cost of mobilization and the safety, or even retrievability, of the equipment when the sampling is completed should be considered. Conditions in the area may suggest that the equipment is on a "one-way-ticket," and therefore expendable. The cost of the equipment should be weighed carefully and the "minimum" taken in for initial stages.

If it is important to drill over water, in streams or rivers, lakes or offshore in bays and open ocean, the size and complexity of the equipment can be formidable problems. An example of such conditions was the drilling with a large Becker drill during the winter months offshore Nome, Alaska, while positioned on ice floes. Offshore jackup barges have also been used with the Becker and with churn drills using rafts and barges.

The mounting of drills on tractor treads can overcome many obstacles on rough terrain.   If the ground is fairly uniform and not too mountainous, it is handy to have the equipment mounted on a truck or on skids and transported to the area by truck, then towed by dozer between locations. However, if these methods are too cumbersome for the terrain, it may be necessary to resort to portable drills that can be carried by helicopter, mule, or by man.

Arctic Regions

In arctic regions where the summer season is limited, speed is of the essence. The cost of moving equipment and operating will be high and is further aggravated by the problem of drilling or digging in permafrost.

Much of the drilling in Alaska by successful dredge mining firms in the past, was conducted in the winter with outside temperatures as low as -35 degrees F. Samples were taken inside enclosed compartments for thawing and panning. The churn drill has been the workhorse in Alaska for many years although the Becker drill has come into limited use, as has the Vibratory drill more recently.

Equipment Comparison

There is no set rule for determining which equipment to use for sampling. However there are some comparisons that can be useful and these are presented in Table A-Placer Sampling Equipment Comparison.

Section 6
Evaluation Procedures

Concentration

The procedure to obtain the final gold sample, which has proven to be most reliable, is using the gold pan/Batea. The sample taken from either a drill core or bulk process may be concentrated to fine black sands using a Rocker. Once it is reduced to a small quantity of fines, an experienced panner can "cut" a sample to such an extent that there will be a high percentage of gold particles (if present), these particles are then estimated.

"Color" Estimation

A standard for preliminary estimation has been developed over the years which is accepted by placer mining engineers and prospectors, that quantifies the weight of gold specs based upon visual observations. These specs may vary in size from a pin head to a pin-point, with four sizes generally being used. Table B Standard To Estimate (Au), Weight In Mgs., Colors No. 1-4, sets forth the basis for estimating weights of gold specs found in field samples.

It is a function of the weight of the sample and based upon a visual estimate, to specify the color number (i.e., 1, 2, 3, or 4). For instance, a #1 color will weigh 3.0 mg. and three #4's will weigh 0.1 mg. Once these estimates are made and the specs counted, that information is written into the log which is described in a subsequent section. The specs are placed in a small glass vial and marked as to the hole and depth of each sample.

Weighing and Preparations

Samples are taken from the vials at campsite after drilling is completed for the day and cleaned. Cleaning is accomplished by using a clean, gold-free globule of mercury about the size of a small bean, placing it into an evaporating dish (three-inch diameter) with the gold specs or particles. The gold will adhere to the mercury and is facilitated by causing a slight rotation of the dish, cleaning any foreign particles from the mercury. This creates a mixture called amalgam.

The globule of amalgam is then placed in either a test tube or other evaporating dish. A diluted solution of nitric acid using 50 ml each of acid and water(1:1) is added using a low heat, as necessary, evaporating the mercury.

NOTE: Care should be taken to cover the tube or dish to prevent breathing any mercury vapors.

Free gold will then be left in a spongelike mass. After the preparation and decanting of the acid, the gold should be carefully washed several times with lukewarm water then dried with a low heat. The sample is now ready for weighing. The gold balance should have an accuracy to 1.0 mg. With extreme care, low gold values can be determined.

Where rusty or coated gold is present, measures should be taken to scour or otherwise brighten the gold prior to attempting amalgamation. This can be done by rubbing it in the gold pan or in a mortar dish with a pestle. The addition of a small amount of lye will counteract oil or grease and will generally assist in amalgamation. The weighed results are then entered into the log as follows.

Log Preparation
The log sheets to be used are normally preprinted and may vary by company as to the tabular format. However certain essential data needs to be included. The elaboration of details about the formation and factors encountered during the drilling process, can be invaluable at later dates when attempting to assess and plan a mining operation and the type of equipment to be specified.

The logs should be filled out in advance with all data and prepared in duplicate for field use. Such data as the drive shoe factor, core factor, and volume factor, which are a function of the drill or method being used, along with either the known or estimated fineness of the insitu gold and weight per unit volume (cubic yards, or cubic meters), should be entered. The samples of logs show both drilling and shafting formats that can be used. As mentioned, the more data that can be included that describe the formation and the nature of the gold, black sands, and other conditions of the strata, the more useful the logs will be at some future date.

It is not unusual to pull logs from files prepared over 60 years ago and study them for an insight to a particular property being considered for prospecting and mining. A complete set of well documented logs may save thousands of dollars in exploration and wrong decisions. It is important that the person filling out the log sign the log and after completing, that it be reviewed either by his supervisor or another placer mining engineer who should then sign it as the "reviewer."

Abbreviations
Table C List of Abbreviations To Be Used On Drill Logs, should be kept available for use in the field. When printing a supply of logs, it is useful to print the abbreviations on the reverse side or have it available in a plastic covered envelope for ready reference. The use of these abbreviations will aid the mining engineers that will be evaluating the logs at future dates.

Corrections & Calculations
On a theoretical basis, when driving a pipe into the ground for sampling purposes, the volume of material forced upward into the pipe should be constant. However, there are a number of conditions that make that volume variable and therefore, in making the calculations, certain corrections must be made.

In order to measure the volume, a measuring stick or box measure system is used into which the sample is poured and measured. If the volume produced in a given drive of, say one or two feet, exceeds the theoretical volume of the pipe for that distance, it is said that the hole has been "over-pumped;" if less than the theoretical then it is "under-pumped." One of the conditions that may cause these variations is friction between the gravel and the drive shoe. Another occurs when encountering rocks or boulders which obstruct the entry of material into the casing. Sometimes an increase in volume may result from the loosening of material from its semi-compact state causing an influx or rise of material into the casing due to hydrostatic pressure.

When there is buried timber it may block the material form entering. However, one of the most important influences on the volume of samples is the length of the drive and a consequent clogging in the casing that may cause the sample to deviate from the theoretical volume.

Using a 4-1/2-inch drill (I.D.) and a 5-1/2-inch cutting shoe, a one-foot drive should produce a sample of 1.73 feet of material in the casing which equals 0.007 cubic yard. A six-inch drill with a 7-1/2-inch cutting shoe would supposedly push into the pipe 1.45 feet of material which equals 0.010875 cubic yard. Working with those examples, correction factors attempt to make allowance for those variables mentioned in the foregoing paragraph to obtain a more accurate figure for entering into the log.

CORRECTION BASED UPON RISE:
Drive (ft)xTheoretical Rise (ft) x Estimated Wt. (mg)/Actual Rise (ft)=Corrected Value

CORRECTION BASED ON VOLUME:
Drive(ft) x Theor. Vol.(cuyd) x Est. Wt. (mg)/Act Vol(cuyd)= Corrected Value
Corrected Value +/- Estimated Weight=Correction where:
Theoretical Rise=Core Factor (given)
Theoretical Volume=Volume Factor (given)
Rise=Core before pumping (-) Core after pumping.

Since the corrected value would be obtained without correction if the ideal amount was pumped out of the pipe, then the estimated weight together with the correction must total the corrected value. Therefore, in making the correction, if the estimated weight is greater than the corrected value, the difference is a minus correction and is positive in the opposite case or a plus correction.

The decision to use the correction based upon RISE or upon VOL-UME depends upon which will result in the more conservative value. In other words, whichever method will reduce or add the least amount of gold to the weight is the method to use. By working with the logs and drill results, the sampler will be able to judge whether the rise or volume basis is more appropriate from a conservative standpoint.

In the cases where rising material tends to be greater than the ideal, it is necessary to combine two or sometimes more drives until the amount of the core after pumping is less than before pumping. These combined drive lengths are then treated as one unit for correction in the usual manner.

All corrections are marked on the log sheets in the space provided for that purpose. If there is no correction, a diagonal line is put in the space to indicate it has received attention. When going through overburden, long drives are the rule and usually result in deficient cores and volumes. When on a long drive and gold is encountered, the correction should be kept to a maximum of 100% of the drive.

Drive Shoe Factor

The drive shoe will be of a diameter that exceeds the inside diameter of the drill casing, using a bit that crowds in the sample to the pipe. The area of the drive shoe must be used to compute the theoretical volume of the sample times the drive length. However, since the total volume crowded into the pipe will not be the theoretical, there must be a factor calculated to multiply times the actual volume to compensate for the added volume.

Table D Standard Values From Drill Hole Data, will provide necessary and convenient factors for calculations. The true volume to be given to a drill hole is determined by substituting for the constant, a factor represented by a corrective factor to offset and eliminate errors due to the variables in the core volumes. In its simplest form, the value of the factor, "F" is expressed by the ratio:
F=Theoretical Core/Actual Core

Actual cores are measured as indicated above, by either recording the linear length of the core which is forced into the casing by each drive or by pumping the material from the hole and measuring it in a calibrated container. Those two measurements are entered on the drill log for each drive and are used in arriving at the value for "F" by comparing them with their theoretical equivalents. The calculation of their equivalents, the derivation and application of the corrective factor "F" are shown as follows.

The amount of material that is driven into the casing is proportional to the effective area, "A" of the drive shoe times the length of the drive. The effective diameter is an important dimension since it affects the value of the drilling constant, "C," because by definition it is:
$C=1 \ cy/A=27cuft/\pi r^2$

Shoe and casing dimensions are usually given in inches, depths in feet and volumes in cubic yards (in metric: cm, m, and m^3, respectively). The value of "C" in feet therefore is expressed as follows:
$C=1 \ cu.yd/A=27 \ ft^3/\pi D^2/4 \times 12"=4950.3/D^2$

The core is usually expanded as it is forced into the casing and experience of years of drilling indicates that this expansion, or "swell" is on an average of 5%. The theoretical rise for any given drive will therefore be:
Theoretical Rise=Drive x D^2/d^2 x 1.05 (d=pipe I.D.)

The theoretical Volume in cubic yards for each drive is as follows:
Volume=Drive(ft)/C x 1.05
Also, since the corrective factor "F" to be applied to the constant "C" is based on:
F=Theoretical Core/Actual Core=Drive x D^2/d^2 x 1.05/Actual Rise
then the corrected value of "C" is C x F.

The value of "C" for a 6-inch drill pipe with a drive shoe diameter of 7-1/2-inch and inside pipe diameter of 5.75-inch, and mean diameter of 6.625-inch has been determined by experience in various materials to be 100. This value is generally accepted and used until a local value is determined that can alter it. From that value of "C," the effective diameter of the shoe or "D" is 7.035-inch or 1.06 times the mean diameter.

On any new property where drilling is undertaken, if we are given the inside and outside diameters of the drive shoe being used, we can calculate a value for "C" that is within the limit of accuracy of the drilling. If we also have the inside diameter of the casing, we can calculate the theoretical Rise and Volume of core from which to determine the corrective factors called for in arriving at the true value of the ground drilled. These dimensions should be noted on the drill logs where different shoes and casing are in use on the same property.

Theoretically, the measured volume recovered should equal the volume corresponding to the equivalent rise. The log entries representing these items are compared and where they are not equal, errors have been introduced which must be corrected. If the measured volume is greater than the volume corresponding to the rise, the correction is determined by the ratio.

The value of "C" varies to some extent with the character of the gravel drilled and, to a lesser extent, with the degree of wear on the drive shoe that may exist prior to changing it. The value of "C" can be closely approximated from the dimensions of the shoe but its exact value, in any given deposit, can only be determined by experience and the careful checking of the drill hole results by other methods.

No correction is usually made for deficient cores. Deficient volumes are generally due to the escape of slimes. They also may be due to the fact that the character of the gravel is such that the pump can drive the core partly out of the drive pipe, or when the water column inside the drive pipe is higher than groundwater level, its weight forces material out through the shoe.

Radford Factor

This factor was derived by William H. Radford, a well known and respected mining engineer in the latter part of the 1800's. While attempting to establish a relationship between drilling and shafting results, Radford sank a shaft 3 feet in diameter using a drill hole as the center, to a depth of 34 feet. The gold obtained from the shaft corresponded almost exactly with the gold from the drill hole when he applied a correction to the drill value with a factor of 0.27.

Noting that the difference in the results you would obtain by varying the correction factor just 0.02 is sufficient sometimes to change the value from net to gross, the factor of 0.27, was therefore established. Thus the Radford Factor has been used for many years as a dependable average, compensating for shoe wear and other variations using a drill.

Final Correction

The weight of gold corresponding to each drive, usually estimated by the panner and entered on the log is multiplied by the correction factor to give the correct weight for that drive. The correction, the amount to be subtracted from the estimated weight, is the estimated weight times(1-F), as shown below:

E = estimated weight of gold
F = correction factor
C = correction to be applied to E
W = corrected weight of gold:

then W= EF and C = E - EF = E (I-F)

The total of the corrections to the various drives for the entire hole is subtracted from the total estimated weight of gold and the correction to the total actual weight of gold is;

Total Actual Corrections=
Total Correction to Estimated Weights x Actual-Wt (-) Repan/Est. Wt.

The corrected weight of gold used for calculation of the final value is:
Final Corrected Wt.=Actual Weight +/- Total Actual Correction

The final corrected weight is used in the following formula:
Value (mg)=gold (oz) x C/Depth (ft).

Table B-Standard Values from drill hole data, gives a few constants for standard churn drill pipe sizes that are, or have been, commonly used.

Computer Program Evaluation(ELAS)

A computer program has been developed by an experienced placer engineering group, ALCON LTD., called ELAS that is tailored for alluvial deposits. The program contains the following functions:

■ Accepts all data of an alluvial drill log and prints a hard copy with internal checks for logic errors.

■ Calculates the logs with various overburden and bedrock cutoffs along with an option of every known system of calculation.

■ Prints clear, usable cross-sections with color coded values using an EPSON LQ-2500 with color kit printer (no plotter needed).

■ Allows the user to select various mining paths with boundary limits to tenths of meters. Either automatic or user-supplied areas for volume calculations, with instantly calculated mineral output.

■ Calculates reserves using different overburden and bedrock cut-offs; Produces a monthly production schedule for any mining/dredge course selected in 4 above.

The Option Supplement

When drilling a borehole, a sample is taken for each interval of drive. The gold content of the interval sample is recorded and the system also applies a gold value correction weight figure for the gold in each interval sample. This information is recorded in the log of each borehole and is entered into the ELAS System database. From the database, this information is used to calculate the mining ore grades of boreholes and of ore reserves.

If you have to alter the gold value correction figures, and recalculate the borehole grades and ore reserve grades, the program offers a number of options. In each event, the gold value correction figure for each interval can be systematically re-estimated, enabling you to calculate the effect on borehole grades and ore reserves, of small systematic changes in the gold values of interval samples.

Although some of the options may look similar on paper, one must take into consideration that the "no positive corrections" factor may be applied to either all the intervals; only overall; or both. In other words, you may choose to apply unlimited corrections to intervals and then reject overall positive corections. It may occur that the sum of the correction weights for each interval is positive; in this example, it would be reset to zero, but again, only at the overall level. When options are enabled (i.e., offered), you are periodically prompted for the option you want. Unlike geological grades, the options will be automatically diabled upon exiting to DOS. The complete program is available from ALCON LTD., GPO Box 2566, Bangkok 10501, Thailand.

Section 7
Phases of Sampling

Theory of "Saturation"
In the late 1800's when churn drills were replacing earlier methods of bulk sampling through shafting, pitting, or trenching, theories were being developed about the various methods of sampling. Later in the early 1900's, projects were being established for dredging where the saturation of drilling of a given property averaged one hole/ten acres or more. But too many variations were occurring after mining the deposits that indicated a higher degree of saturation was needed.

Mining properties were supposedly validated but after bringing in a dredge, the values were less than the estimated quantities. This led to testing with shafts to attempt to correlate drill results with the old "dependable" means of manual labor and such relationships as the "RADFORD FACTOR," led some to believe that drill holes should be in a similar distribution.

However, more extensive drilling of properties showed this to be an incorrect premise. This resulted in placer miners increasing the saturation. Some prospects were proven to less than one acre-per-hole, depending upon the overall tenor and other factors that indicated random values.

As this experience proceeded with churn drilling being verified with actual dredging, another phenomenon began to develop; greater values dredged than indicated by drilling. This was particularly evident after some of the more progressive dredging companies, such as PATO CONSOLIDATED GOLD DREDGING, YUBA CON-SOLIDATED GOLD FIELDS, NATOMAS CO., and U.S. SMELT-ING, REFINING & MINING CO., began to introduce the mineral jig.

The R/E (Recovery divided by Exploration or drill results) not only became close to unity but exceeded it. Evidence then began to show that there was a point at which the saturation of a property by drilling could decrease the average apparent value of the ground which led to the term "oversaturation" or more holes drilled than some optimum point. From this a ratio of between one and five acres to one drill hole, became a more accepted rule.

Reconnaissance

The degree to which reconnaissance of a prospect can be useful may vary in proportion to the background experience in placer evaluation and mining by the field investigator. Likewise, that experience can be invaluable in the phase of literature and data search on the prospect before making a field visit. It should be clear that inexperienced personnel should not be relied upon for decisions and fact-gathering at this stage unless under the direct supervision of a senior placer mining engineer.

The following provides a check list for field observations that can be useful. The resultant data becomes a building block for the file and planning for future follow-on sampling that may be conducted. Some of the important physical aspects to be observed are indicated:

■ General topography, drainage, physiography and geology.

■ Previous or existing mining, pitting, trenching and other physical evidence including early activities of Chinese miners.

■ Bedrock outcroppings related to volumetric estimates and conditions for mining including hardness sampling, irregularities and similar anomalies that might relate to subsequent dredging.

■ Gravel deposits or sand bars that might indicate a large volume as a basis for volumetric estimates as well as sizing of boulders.

■ Presence of clay and cemented strata that might be visible in cliffs or terraces, which would have an influence on mining efficiency and recovery.

■ Evidence of hidden paleochannels and terraces sometimes related to feeder streams or river tributaries.

■ Ecological or environmental conditions that might be a problem in gaining permits for mining, such as farmland, fisheries, reservoirs, aquifers, forestry, etc,

■ Hydrological conditions to be considered; the shortage of water or too much water, river hydraulics and diversion problems, berms, dikes, flood protection. Early indications of problem areas can be valuable in assessing potential risk factors.

Surface Samples
Wherever an alluvial deposit may have subsurface exposure such as from earlier pitting or shafting, it is useful to gather samples from the bottom or sides and pan them on site. Sometimes a stream may produce evidence of gold on the shoreline and should also be panned to aid in decisions about taking a subsequent step to drill or bulk sample a property. The purpose of taking surface samples during the Reconnaissance phase is to establish if there is a presence of gold, and to aid in the planning for a subsequent phase.

Scout Drilling/Sampling
The goal of this initial phase of subsurface sampling is to determine the occurrence and quantity of gold at various levels and to sample bedrock which is normally the dominant location of alluvial gold concentrations. It is important to determine, for instance, where early channels may be located and therefore a line of holes across the river or stream basin can help to identify that location.

This can be difficult when working in large alluvial deposits such as in the desert, flood plains, or elevated terraces where water is not flowing, nor indications of early flows which may be buried without surface evidence. A study of the geomorphology may become of such complexity that a professional in that field should be consulted (see Appendix-Case Study B-Nechi).

The location of the channel bottom is an important factor for an evaluation in the scouting phase. Additional holes may be important to delineate economic boundaries or limits of the deposit. Once those facts have been established, it should permit rough calculations of estimated total volume of ground and tenor or Au (mg/cy) for a preliminary feasibility analysis of a mining operation.

Selection of Equipment and Method of Sampling
Some of the factors that can lead to a decision as to what approach to sampling should be used during the Scouting phase include: Depth of

ground to probable bedrock and presence of boulders or hard material, can make drilling difficult and time consuming.

With a trend to seek deeper deposits or "deep leads" including tertiary gravels as deep as 1,000 feet, few of the earlier, established methods of sampling are relevant. Some mining groups use rotary drills, which can be used to penetrate hard materials followed by the driving of a casing into the last part of the formation to bedrock.

This may be sufficient for the groundwork to establish the final method of mining (which is not a clear issue at greater depths). The caisson drill has been used for greater depths (150-300 feet). If the ground is shallow, it may be enough to dig a shaft and channel sample or to use a backhoe for the bulk sample since both methods are inexpensive to accomplish. We recommend, however, the use of a churn drill wherever possible in view of the assurance of obtaining an undisturbed sample, which is difficult in the case of bulk sampling.

Terrain

The type of method or equipment selected will also be greatly affected by the type of terrain or environmental conditions. A water-covered environment will eliminate a number of methods and may force the use of some of the earliest tools such as the Banka drill.

The opposite may prevail when enough evidence suggests that the probabilities are high for finding gold and there is sufficient budget to permit bringing in a faster, more expensive drill. In fact, when the speed of larger drills such as the Becker are considered in comparison with labor and engineering costs, as well as the benefits of expediting results, it is often a lower total cost to employ than the slower churn drills. These and other factors must be weighed to make a final decision that will be both economical and reliable.

Spacing of Holes

As mentioned above, the nature of the deposit on the surface will often dictate the approach to its sampling. If it is a well-defined river or stream valley with terraces easily identifiable, a line of holes initially spaced on 100-200 feet centers depending on the width of the alluvial bed, could be drilled initially. If the location of the channel becomes apparent, additional holes should be drilled in between the original holes until it is well located and bedrock tested for values.

Once the boundaries of the flow are determined, wider spaced holes should be drilled up river to establish limits of the deposit and variations in grade, if any. This may be automatically limited by the size of the property but it is important for determining total volumes and potential of the mine.

The approach to evaluation can be to try and locate the upper limits of the flow and thus the highest values, followed by searching for the lower limits at the other extreme. This may depend on ownership or other indirect factors of the property that could affect economic results but may lead to a decision to drill from the lowest portion of the property with the philosophy that, if the values there are too low, it would so degrade the total volume of the prospect that it would eliminate the prospect as being economic. In some cases, a scout line across a large flood plain may be as long as five to seven miles.

Indicated Reserves

When the results of the Scout Drilling/Sampling phase is completed and the results are favorable, the next step is to plan a more extensive and well-ordered grid drilling program. With today's cost of money and labor, the use of bulk sampling, unless it is with an engine-powered caisson, is usually not efficient. As has been mentioned in our discussion of methods, there are accuracy problems with bulk sampling and depth limitations when using manpower or pitting with backhoes.

Therefore, we will consider primarily the mechanized methods of sampling. Considering that this must be done in terms of drilling parallel lines over the extent of the deposit, time becomes an important factor. It is not unusual to mobilize up to eight or more churn drills to saturate a large property and to spend several years doing it. Time however, is too costly today in most cases and decisions are demanded in shorter periods. The use of higher speed drills almost becomes mandatory unless the deposit is so shallow or soft that hand or Keystone-type churn drills will do the job. The Banka or comparable hand drill, using low cost native labor is an alternative in third world countries.

The purpose of the exploration phase is to determine with a grid system guided by the scout drilling results as to location of channels, limits of the deposit and a more reliable estimate of the consistency of the property's values. When this phase is completed, the resulting reserves, if favorable, may be termed INDICATED RESERVES. The results should provide a basis for a more definitive feasibility study and on which preliminary financing decisions and solicitations can be based.

Equipment Selection
The same rationale for selecting of sampling equipment for Scout Drilling may apply. However, a larger prospect for exploration requires a different magnitude of decisions affecting budget, size of prospect, importance of accelerating results and physical conditions. Since placer deposits are found in virtually every climate in the world, from the equator to the arctic, conditions can vary accordingly.

In the frozen climes, permafrost and low winter temperatures must be dealt with. In the tropics, swamps, jungle diseases and monsoon seasons must be accommodated. Once the above and other applicable factors have been assessed, it may prove to be more economic to buy a new drill(s) in view of the extent of the exploration and values concerned.

Grid Spacing & Saturation

No set rule can be stated for the spacing of lines or holes to establish INDICATED RESERVES in the exploratory phase, in view of the different physical conditions or size of properties. Even saturation rules are variable but a ratio of one hole-per-ten acres could be considered useful in final calculations of the number of holes to drill in planning the grid and drilling plan.

It should be noted, however, that should a bulk sampling program be used where either shafts, pits, trenches or caisson holes are sunk, a different computation would be made as to the number required. Obviously, one three-foot diameter caisson sample in ten acres would have a higher influence than a single, five-inch drill hole.

The nature of the subsoil of the deposit would have a bearing upon a decision of necessary saturation and requires the judgement of an experienced placer mining engineer to decide. It is also important to point out the variability of the word EXPLORATION, since it implies more than a fixed connotation to different engineers or geologists and is strongly influenced by the property characteristics and magnitude.

Where there is a concession 100 miles in length, the initial exploration drilling after scout drilling may make a line of holes one to ten miles apart. This will mainly be used to identify the best prospect zones and would be followed by a selection of most favorable or high priority areas with exploratory drilling using closer lines such as every 1,000 feet. It should also be mentioned that in many cases, as a result of earlier drilling or sampling being conducted on a given property where the results are available, exploratory drilling may serve the purpose of verifying those results.

In that case, it is appropriate to sink holes or shafts in close proximity to previously recorded holes, picking highs and lows, to see if those value ranges can be verified with new drill or shaft holes. Should a level of confidence develop in the earlier sampling results, then a rationale can be developed to merely increase the saturation by adding to the earlier results with the verification drilling and adopting a saturation level at some higher point.

Development Drilling: Proven Reserves

When the exploratory drilling phase has been completed and the results have been considered sufficiently economic to proceed with the final phase of drilling, the grid spacing is increased in density by interspersing parallel lines and the number of holes-per-line.

Depending upon the tenor of the deposit, this may influence a decision relative to saturation of the property with drilling or other sampling methods. Normally, the requirement at this stage for proper verification of reserves will be with a drill. If the tenors are high (i.e., for a placer property any values above 200 mg/cuyd would be considered as high), then the existence of INDICATED RESERVES would tend to establish a reasonable level of confidence. In those circumstances, in a large deposit a ratio of three to four acres to one hole could prove to be adequate. In a smaller deposit where there is less margin for error, you would keep to a higher ratio and nearer to one acre-per-hole.

Equipment Selection

The drilling equipment selected for the exploratory phase would normally be appropriate to the development drilling phase. However, depending upon the scope or size of the prospect, additional drills could be added at this stage to accelerate completion of the project. For instance, one Keystone drill may have been used for exploration. In the development phase either multiple churn drills or a larger and faster system, such as the Becker drill, could be brought in to compress the time of completion.

If a Becker Drill were chosen, it is important that a series of tests be run to insure that there is a dependable correlation established between the Becker and the previous drilling method. This could result in a revised correction factor being calculated which, when applied to the

Becker results, would correlate closely to the Keystone values. As with exploration, the use of Bulk Sampling techniques is not normally considered a suitable means of validating Proven Reserves.

Section 8
Physical Considerations of Prospects

In an effort to give the broadest possible discussion of conditions and the methods that experience has shown to be reasonable, in approaching each of them, the following paragraphs will present some of the main categories and cite some of our experience with their unique problems in placer evaluation.

Rivers and Streams
The most common source of placer gold occurrence is, by definition, in river and stream beds. That will include their adjacent terraces and earlier, buried paleo channels. Whether these areas have been confined by hills and mountains or are in a broad flood plain, can determine the extent of the deposit and the method of approaching its evaluation.

Typical broad flood plains are in the tropics in such areas as South America, SE Asia and Africa. Some have produced very large placer deposits which are being dredge mined to this day and continuously since the beginning of this century. Areas such as these require extensive exploration programs to properly assess and can be risky as in any area, if short-cuts are taken.

In such areas it is not unusual to have a single line of drill holes extend for several miles to cover the total area of potential reserves. Drilling crews used to spend a year just completing one line, but today it is difficult to think in terms of long range drilling of that nature. Faster methods are now available as indicated and should be used when planning such a program. As long as the ground is relatively dry, tractor or skid-mounted drills can be used. If in swamps or over running water, canoes, tethered rafts, barges or jackup barges must be used depending upon the size of drill.

Ocean Regions

Most gold mining of placers involved with the sea have been in the immediate litoral zones or elevated beaches. Some attempts have been made to prospect properties offshore with an idea of finding an economically feasible way of mining them. Such was the case in offshore Nome, Alaska, which was extensively drilled in the mid to late 1960's, using a Becker Hammer R.C. drill during the winter from the ice. The summer water conditions with a high frequency of storms makes it difficult and expensive to dredge.

In the late 1980's, WESTGOLD brought in a large, BL/M 30 ft^3 dredge with 50 meters digging depth, from SE Asia that had been used for tin mining off Banka Island, Indonesia. Named the "BIMA," this is the largest dredge of that type built and was designed for open ocean operations with a 10-foot freeboard. They modified the recovery plant to recover gold and operated offshore Nome for a few years, but did not find it economical, mainly due to the high cost of operation of the dredge, and shut the project down.

Other prospects for offshore gold deposits do exist but have not been considered economical. Mineral prospects including phosphates, rutile, magnetite, ilmenite, zircon and diamond are the most promising. There is current mining of such alluvial deposits and the potential for the future looks promising.

Tertiary Gravels

A great deal of prospecting and evaluating has been done in recent years in some of the deposits, attempting to find a means to mine them. They represent deep deposits of gravel from earlier periods that were the subject of hydraulicking in the early gold rush days of the mid-

1800's.   Since those practises were outlawed for environmental reasons in the late 1800's, and is a wasteful method of mining with low recoveries in any event, new ideas of mining have been explored.

The deep gravel deposits of Northern California, New Zealand and Australia are particularly well known.   Mining underground with pillar and room methods, has been evaluated.   This mining approach has lent itself in such areas as Siberia with "Gulag" labor, but is very severe.   Open pit mining is another method that has been examined. There are large deposits available if mechanized methods and humane conditions can be devised.

Moraine Deposits
There are some outstanding placer gold deposits laid down by moraine such as in the South American Andes, Alaska and Canada.   There is some debate among geologists as to the origin; whether alluvial and moraine combination.   Some of these deposits tend to be too deep for most alluvial dredging systems.

The best example we know was the project in Peru by NATOMAS in the 1960's.   They drilled the deposit and proved up sufficient reserves with a depth of 40 feet, even though the gold occured deeper.   Budget restraints of the time due to the fixed gold price of $35./oz, caused them to compromise on a small, used dredge. After operating at 17,000 feet elevation, turning the dredge over a couple of times due to high center of gravity, the project was stopped in 1971.

Attempts to date by the government who took over the property, to restart the operation have failed for lack of financing (see Figure P13.2).

Desert Placers

Prospecting for placer deposits in desert regions has had an appeal to prospectors for many years. The presence of "flood gold," caused by periodic cloud bursts and sudden flows of water runoff, scatters fine gold at the surface. Seldom however, does it occur in depth or economic quantities but nevertheless is the source of many mining scams.

Deserts have produced substantial, hardrock gold mines, and thus has encouraged placer prospecting. Few have resulted in commercial scale placer mines, however. The shortage of water often discourages placer mine development, but is minimized by the promoters or suggested that "dry recovery" techniques will work. None have resulted in commercial-scale mines to our knowledge.

Some of the best prospects in the desert have proven to be Eluvial deposits; the result of weathering of hardrock deposits that have decomposed and released the gold in situ. This type of deposit requires knowledgeable evaluation by placer mining engineers and geologists, in order to relate the deposit to mining. The usual alluvial deposit characteristics, for instance, are not present. The gold will often occur from surface to some point of cutoff, but no bedrock will exist.

Section 9
Becker Drill Tests

Introduction

Drilling was conducted over a period of some 50 years, using Keystone-type drills (i.e., churn drills), on the Taramakau and Grey Rivers, West Coast, South Island of New Zealand. In order to prove up large volume reserves, a Becker Reverse Circulation (RC), Hammer drill was purchased and shipped to New Zealand in 1981. However, at that time there had been no verification of results using the Becker on a commercial-scale, placer gold dredging operation. The Becker had been used by our company on a number of placer prospects and shown to be able to drill at a speed approximating 10 times that of a standard churn drill.

A project of drilling a large deposit offshore Nome, Alaska, using a Becker drill during the winter months from ice floes, was accomplished by ASARCO in 1968-70. However, no verification of the deposit had yet been made with a dredge or other means. CPD brought in Joseph Wojcik, Consulting Mining Geologist who had been involved with the Nome drilling for ASARCO, to work with our mining engineers to determine an adjustment factor for the Becker Drill. The deposit on the Grey River and another nearby on the Taramakau River, had been drilled extensively by churn drills in previous years. A 20 cuft BL/M dredge was operating on the latter river and producing gold.

When the Becker drill was first introduced we began drilling on the Taramakau, attempting to extend the reserves ahead of the dredge. Results of the first correlation attempts with the churn drill results were erratic. Therefore, there was a need to establish the reliability of the Becker samples which we felt could best be done in relationship to the well established and proven, churn drilling.

Test Procedures

The first attempts to obtain correlations between Becker and churn/Keystone drill holes, included drilling three Becker holes in a circle around one Keystone hole. The weighted averages were compared in that way at five Keystone holes on the Taramakau River. The same tests were subsequently made at 12 Keystone drill sites on the Grey River.

In addition to the circling of Keystone holes, two rectangular blocks were selected on the Grey River, for which ore reserve estimates were calculated from drill results. Both Keystone and Becker drill holes were completed in the area. The tenor of the Keystone ore reserve was 1.35 to 1.51 times the tenor of the Becker ore reserve, on the first calculations. With an adjustment in the factors used in calculating values from Becker drilled holes, the tenor of the Keystone ore reserve was revised to a range of 0.96-1.02 times the tenor of the comparable Becker ore reserve calculations.

Having arrived at essentially the same ratios from four different sets of data, it was determined that the Becker values calculated with the new adjustment factors can be compared directly to the Keystone values. Comparison of the size distribution of gold recovered from Keystone and Becker drilled holes reinforces this confidence. Results of the drilling comparisons are included in the following Table 1.

Size Distribution

The total of the color counts in each of the four size ranges was added for 28 Keystone holes and 81 Becker holes; the total counts being multiplied by the appropriate weight to arrive at estimated weights. The results show that gold is not lost selectively by the Becker drill. Figure 1 shows the percent size distribution and Figure 4 is a screen size analysis of gold from Taramakau River drill holes, plotted on the same curve.

PERCENT DISTRIBUTION
BECKER FACTORS

A plot of the percent distribution of measured volumes of samples recovered from five-foot intervals in Becker holes, has a mode at about 0.275 cuft, which is 1.38 times the theoretical volume expected from a core of 2.7 inches diameter. The drive shoe factor for a 2.70-inch-core is 675. The second Becker calculations reported in Table 1, used a theoretical loose volume of 0.275 cuft-per-foot of interval and a drive shoe factor of 675.

Conclusion

A consistent correlation of results from Becker drilled holes with results from Keystone drilled holes on the Grey River, can be achieved by allowing for expansion of the in-place sample into the loose measured volume and for a core diameter of 2.70 inches. All of the evidence compiled from 121 Becker holes and over 58 Keystone holes, indicate that the Becker drilling technique and equipment, if properly monitored, provides results that may be merged on an equal basis with Keystone results.

Linear Regression

To define this relationship further, we compared the values of the 12 sets of three Becker to one Keystone holes on the Grey River. Using least squares linear regression, calculated the best fit curve for the pairs of values and the correlation coefficient between the sets of pairs. Two of the holes, DEK8 and 11CK, appeared on the graph to be either erratically low or high, so best fit curves and correlation coefficients were calculated also for the ten pairs of values without these two holes. The relationship between Keystone and Becker values for the 12 sets was 1.6:1, using hole tenor only; and 1.54:1 using the depth x grade product. This ratio was unchanged when the two erratic holes were deleted. The correlation coefficient improved from 0.68 to 0.94, however, when the erratic two holes were deleted.

On the Taramakau River, 5 sets of 3 Becker to 1 Keystone holes were drilled and the ratio of the average grades was 1.5 Keystone to 1 Becker, yet the best fit linear regression curve has a slope of 1.1:1. The correlation coefficient is only 0.41 so there are not enough sample sets from which to draw conclusions.

Core Rise

Since the corrections in the individual intervals in the drill holes are based on the volume of sample recovered, and the overall tenor is calculated from the theoretical core size, we wondered if one or both of these parameters were contributing to an over-discounted value calculated from the Becker samples.

We measured the core rise in 58 drives on the Becker in six holes and measured the volume recovered in the drives in four of these holes. Core rise was measured by lowering a weighted measuring tape inside the pipe before a drive and noting the length of open tubing. The pipe was then driven either 12 inches or 6 inches without air or water circulation and the length of open tubing measured once more. The amount of core incorporated into the tubing was the difference between the before and after measurements. After the second measurement, air and water were circulated to bring the sample to the surface. Three or four one-foot drives were measured in each five foot interval, but the volume for the five foot interval was measured jointly.

The average rise for 51 measurements of 12-inch drives was 9.36 inches, and for 7 measurements of 6-inch drives, was 5.16 inches; an overall 78% of theoretical in-place volume. Volume recovered from these same intervals was 102% of the theoretical in place volume of a 3.125-inch diameter core. Was there expansion of the 78% of a 3.125-inch-core when it was displaced through the tubing into the volume measuring bucket or was there 100% of a 2.70-inch-core incorporated into 78% of the equivalent length of 3.125 inches tubing and then expanded in the volume bucket?

Fifty-one intervals in seven Keystone holes were analysed, in which intervals of obviously rising core, were excluded. Core rise averaged 89% of theoretical using a factor of 1.4-foot rise for 1-foot advance, and the measured volume was 98% of theoretical using the "Radford factor" of 0.27 cuft/ft of advance, but when the measured volume was compared to the throretical volume removed by the sand pump from the 5.75-inch ID casing, the loose volume was 38% greater than theoretical.

The volume removed was calculated from the difference in the height of core inside the pipe before and after pumping, times the area of the pipe. Checking this further, we drove a section of Keystone pipe with the Becker, pulled it out, measured the length of core retained inside then emptied it into the volume bucket for measurement. In two tests, the core rises were 58% and 55% of theoretical and the volumes were 144% and 140% of theoretical. Certainly some compaction takes place in driving the casing ahead but to what degree is still in question. That the drilled loose volume is expanded in respect to the in-place volume is demonstrated.

Ore Reserves
Zone A selected for comparison of drill hole samples by calculation of ore reserves, was evaluated four different ways. First, an estimate of cubic yards and grade (1) was calculated using only Keystone holes. Triangles were constructed by drawing lines between holes on the map. Where there was a choice of two or more triangles, the one chosen was the most nearly equilateral. Average depths for each triangle were the arithmetical average of the depths of the holes at the vertices.

Average grade was the sum of the depth time grade products for the bounding holes, divided by the sum of the depths. Around the boundaries of the zone, rectangles were constructed by drawing lines from the holes perpendicular to the side lines. Depths and weighted grades were calculated for the rectangles using such holes as were available at the corners. Areas for the triangles were calculated from the coordinates of the drill holes and areas for the rectangles were calculated from dimensions scaled from the 1:7920 plan map. Areas x depths gave volumes for the individual polygons and sum of the grade x volume products, were divided by the sum of the volumes to derive the weighted average grade.

A second estimate (2) of cubic yards and grade was done similarly, using only Becker drill holes with the grades calculated by the formula (corrected gold weight x 7.825) divided by depth=gr/cy; the corrected weight calculated by the formula actual weight-(sum of corrections x actual wt)-repan=corrected weight and the individual corrections determined by the measured volume using a theoretical volume of 0.053 cf/ft of hole.

The third estimate of cy and grade (3) was calculated similarly to the first using only Becker holes with the grades calculated using the same formula as in (2); the corrected wt calculated by the formula actual wt-(sum of corrections x actual wt/est wt)=corrected wt and the individual corrections determined by measured volume, using a theoretical volume of 138% of 0.053 or 0.073cf/ft.

The fourth estimate of cy and grade (4) was calculated similarly to the first, using only Becker holes with the grades calculated by the formula (corr Au wt x 10.42)/depth, the corrected wt calculated as in (3), and individual corrections determined by the measured volume using a theoretical volume of 0.053 cf/ft.

Thus, estimate #2 represents the results assuming a 3.125-inch diameter core and no expansion into the loose volume. Est #3, represents the results assuming a 3.125-inch diameter core and 38% expansion into the loose volume. Est #4 represents the results assuming a 2.70-inch diameter core and 33% expansion into the loose volume.

The volumes, grades and relationships to the Keystone estimate are tabulated in Table 1. The Keystone/Becker ratios are compared to the ratios between the 3:1 drill hole sets on the linear regression curves in Figures 5 and 6, using Becker hole grades, calculated in the same manner as est's 2 and 3. Arriving at the same ratios in the 3:1 drill holes and in the est's of cy's and grade shows that the Becker samples can be used reliably; that a core diameter less than the actual shoe ID, is probably being recovered and that some expansion of the sample into the loose volume does occur.

A distribution of measured volumes of all five-foot sample intervals in 79 Becker holes (Figure 7) shows the mode at about 0.275 cf/5-foot. This is equivalent to a volume of 0.055 cf/ft which is 38% expansion of a 2.70-inch diameter core. If the Becker grades are calculated using these theoretical parameters, the results should be readily comparable to Keystone results.

Cross-Sections
A visual comparison of Becker and Keystone sampling is available on cross sections where the gold content in an interval, is plotted as a horizontal bar graph, in mg/ft. The Keystone values are plotted at 1/5 of actual to keep them on the same scale as the Becker, which recovers 1/5 of the quantity of sample of the Keystone. A sample cross-section is included as Figure 8, and the consistency between is readily apparent.

Conclusions

Volume and grade estimates calculated for Zone A, and comparison of gold size distribution in samples from the Grey River, show that the Becker samples may be used with confidence to evaluate the tenor of the ground.

Apparently, some effect of the drilling technique, the shoe configuration, or the core size, causes the loose volume to be greater than the in-place core volume. When that expansion is taken into account, the Becker values correspond very closely to Keystone values in the same area. The calculated correlation co-efficients of 0.95 are very close to the perfect 1.0.

The Becker holes averaged 1.5-foot deeper than the Keystone holes, which estimated a slightly greater volume at a slightly lower grade. Total estimated gold contained in the 110 acre Zone A, was only 1.7% higher in the Keystone calculation than in the Becker #3 and 2.6% higher than in the Becker #4 calculation.

Use of a theoretical 2.7-inch diameter and a 0.055 cf loose volume/feet of core in future Becker calculations, should allow the Becker and Keystone values to be merged with no other special treatment. Further drilling should be integrated with a least one Keystone hole to each five Becker holes and the relationship monitored; so that we may be alert to possible changes in Becker parameters due to changing ground or drilling conditions.

Basic Placer Mining

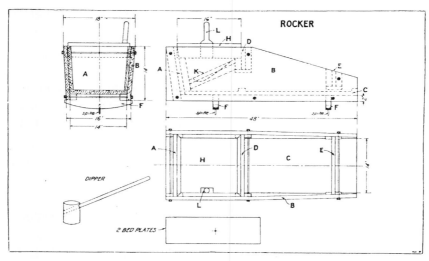

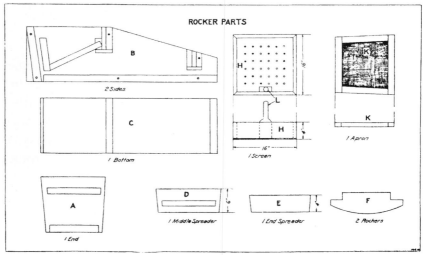

FIGURE 1-ROCKER SCHEMATIC DRAWING

TABLE A

PLACER SAMPLING EQUIPMENT COMPARISON

Type	Capital[1] Cost-$US	Penetration Ft./10 Hrs.	Number in Crew Drilling	Number in Crew Sampling	Manhours Per Ft.	Maximum Depth-Ft.
DRILLING						
Banka Drill	$15,000	20	5	2	3.5	50
Ward Drill	8,000	10	12	2	14.0	50
"Airplane"	50,000	25	4	2	2.4	75
Keystone	75,000	30	2	2	1.3	100
Becker, Hammer	400,000	200	2	3	0.25	150
BULK SAMPLING						
Shafting	NA	5	3	2	10.0	30
Backhoe, Pitting	Rental	20	2	3	2.5	20
Caisson, Power	500,000	50	4	3	1.4	200

NOTE: ABOVE DATA ASSUMES SAME MATERIAL: CLEAN GRAVELS.

[1]Relative range of prices - 1983.

TABLE A-PLACER SAMPLING EQUIPMENT COMPARISON

TABLE B

STANDARD TO ESTIMATE (Au)
Weight in Mgrs.

NO. 1 COLORS

Colors	0	1	2	3	4	5	6	7	8	9
		3.0	6.0	9.0	12.0	15.0	18.0	21.0	24.0	27.0
10	30.0	33.0	36.0	39.0	42.0	45.0	48.0	51.0	54.0	57.0
20	60.0	63.0	66.0	69.0	72.0	75.0	78.0	81.0	84.0	87.0
30	90.0	93.0	96.0	99.0	102.0	105.0	108.0	111.0	114.0	117.0
40	120.0	123.0	126.0	129.0	132.0	135.0	138.0	141.0	144.0	147.0
50	150.0	153.0	156.0	159.0	162.0	165.0	168.0	171.0	174.0	177.0
60	180.0	183.0	186.0	189.0	192.0	195.0	198.0	201.0	204.0	207.0
70	210.0	213.0	216.0	219.0	222.0	225.0	228.0	231.0	234.0	237.0
80	240.0	243.0	246.0	249.0	252.0	255.0	258.0	261.0	264.0	267.0
90	270.0	273.0	276.0	279.0	282.0	285.0	288.0	291.0	294.0	297.0
100	300.0	303.0	306.0	309.0	312.0	315.0	318.0	321.0	324.0	327.0
110	330.0	333.0	336.0	339.0	342.0	345.0	348.0	351.0	354.0	357.0
120	360.0	363.0	366.0	369.0	372.0	375.0	378.0	381.0	384.0	387.0
130	390.0	393.0	396.0	399.0	402.0	405.0	408.0	411.0	414.0	417.0
140	420.0	423.0	426.0	429.0	432.0	435.0	438.0	441.0	444.0	447.0
150	450.0	453.0	456.0	459.0	462.0	465.0	468.0	471.0	474.0	477.0
160	480.0	483.0	486.0	489.0	492.0	495.0	498.0	501.0	504.0	507.0
170	510.0	513.0	516.0	519.0	522.0	525.0	528.0	531.0	534.0	537.0
180	540.0	543.0	546.0	549.0	552.0	555.0	558.0	561.0	564.0	567.0
190	570.0	573.0	576.0	579.0	582.0	585.0	588.0	591.0	594.0	597.0
200	600.0	603.0	606.0	609.0	612.0	615.0	618.0	621.0	624.0	627.0
210	630.0	633.0	636.0	639.0	642.0	645.0	648.0	651.0	654.0	657.0
220	660.0	663.0	666.0	669.0	672.0	675.0	678.0	681.0	684.0	687.0
230	690.0	693.0	696.0	699.0	702.0	705.0	708.0	711.0	714.0	717.0
240	720.0	723.0	726.0	729.0	732.0	735.0	738.0	741.0	744.0	747.0
250	750.0	753.0	756.0	759.0	762.0	765.0	768.0	771.0	774.0	777.0
260	780.0	783.0	786.0	789.0	792.0	795.0	798.0	801.0	804.0	807.0
270	810.0	813.0	816.0	819.0	822.0	825.0	828.0	831.0	834.0	837.0
280	840.0	843.0	846.0	849.0	852.0	855.0	858.0	861.0	864.0	867.0
290	870.0	873.0	876.0	879.0	882.0	885.0	888.0	891.0	894.0	897.0
300	900.0	903.0	906.0	909.0	912.0	915.0	918.0	921.0	924.0	927.0
310	930.0	933.0	936.0	939.0	942.0	945.0	948.0	951.0	954.0	957.0
320	960.0	963.0	966.0	969.0	972.0	975.0	978.0	981.0	984.0	987.0
330	990.0	993.0	996.0	999.0	1002.0	1005.0	1008.0	1011.0	1014.0	1017.0
340	1020.0	1023.0	1026.0	1029.0	1032.0	1035.0	1038.0	1041.0	1044.0	1047.0
350	1050.0	1053.0	1056.0	1059.0	1062.0	1065.0	1068.0	1071.0	1074.0	1077.0
360	1080.0	1083.0	1086.0	1089.0	1092.0	1095.0	1098.0	1101.0	1104.0	1107.0
370	1110.0	1113.0	1116.0	1119.0	1122.0	1125.0	1128.0	1131.0	1134.0	1137.0
380	1140.0	1143.0	1146.0	1149.0	1152.0	1155.0	1158.0	1161.0	1164.0	1167.0
390	1170.0	1173.0	1176.0	1179.0	1182.0	1185.0	1188.0	1191.0	1194.0	1197.0
400	1200.0	1203.0	1206.0	1209.0	1212.0	1215.0	1218.0	1221.0	1224.0	1227.0

TABLE B-STANDARD TO ESTIMATE (Au): NO. 1 COLORS

NO. 2 COLORS

Weight in Mgrs.

Colors	0	1	2	3	4	5	6	7	8	9
		1.3	2.6	3.9	5.2	6.5	7.8	9.1	10.4	11.7
10	13.0	14.3	15.6	16.9	18.2	19.5	20.8	22.1	23.4	24.7
20	26.0	27.3	28.6	29.9	31.2	32.5	33.8	35.1	36.4	37.7
30	39.0	40.3	41.6	42.9	44.2	45.5	46.8	48.1	49.4	50.7
40	52.0	53.3	54.6	55.9	57.2	58.5	59.8	61.1	62.4	63.7
50	65.0	66.3	67.6	68.9	70.2	71.5	72.8	74.1	75.4	76.7
60	78.0	79.3	80.6	81.9	83.2	84.5	85.8	87.1	88.4	89.7
70	91.0	92.3	93.6	94.9	96.2	97.5	98.8	100.1	101.4	102.7
80	104.0	105.3	106.6	107.9	109.2	110.5	111.8	113.1	114.4	115.7
90	117.0	118.3	119.6	120.9	122.2	123.5	124.8	126.1	127.4	128.7
100	130.0	131.3	132.6	133.9	135.2	136.5	137.8	139.1	140.4	141.7
110	143.0	144.3	145.6	146.9	148.2	149.5	150.8	152.1	153.4	154.7
120	156.0	157.3	158.6	159.9	161.2	162.5	163.8	165.1	166.4	167.7
130	169.0	170.3	171.6	172.9	174.2	175.5	176.8	178.1	179.4	180.7
140	182.0	183.3	184.6	185.9	187.2	188.5	189.8	191.1	192.4	193.7
150	195.0	196.3	197.6	198.9	200.2	201.5	202.8	204.1	205.4	206.7
160	208.0	209.3	210.6	211.9	213.2	214.5	215.8	217.1	218.4	219.7
170	221.0	222.3	223.6	224.9	226.2	227.5	228.8	230.1	231.4	232.7
180	234.0	235.3	236.6	237.9	239.2	240.5	241.8	243.1	244.4	245.7
190	247.0	248.3	249.6	250.9	252.2	253.5	254.8	256.1	257.4	258.7
200	260.0	261.3	262.6	263.9	265.2	266.5	267.8	269.1	270.4	271.7
210	273.0	275.3	275.6	276.9	278.2	279.5	280.8	282.1	283.4	284.7
220	286.0	287.3	288.6	289.9	291.2	292.5	293.8	295.1	296.4	297.7
230	299.0	300.3	301.6	302.9	304.2	305.5	306.8	308.1	309.4	310.7
240	312.0	313.3	314.6	315.9	317.2	318.5	319.8	321.1	322.4	323.7
250	325.0	326.3	327.6	328.9	330.2	331.5	332.8	334.1	335.4	336.7
260	338.0	339.3	340.6	341.0	343.2	344.5	345.8	347.1	348.4	349.7
270	351.0	352.3	353.6	354.9	356.2	357.5	358.8	360.1	361.4	362.7
280	364.0	365.3	366.6	367.9	369.2	370.5	371.8	373.1	374.4	375.7
290	377.0	378.3	379.6	380.9	382.2	383.5	384.8	386.1	387.4	388.7
300	390.0	391.3	392.6	393.9	395.2	396.5	397.8	399.1	400.4	401.7
310	403.0	404.3	405.6	406.9	408.2	409.5	410.8	412.1	413.4	414.7
320	416.0	417.3	418.6	419.9	421.2	422.5	423.8	425.1	426.4	427.7
330	429.0	430.3	431.6	432.9	434.2	435.5	436.8	438.1	439.4	440.7
340	442.0	443.3	444.6	445.9	447.2	448.5	449.8	451.1	452.4	453.7
350	455.0	456.3	457.6	458.9	460.2	461.5	462.8	464.1	465.4	466.7
360	468.0	469.3	470.6	471.9	473.2	474.5	475.8	477.1	478.4	479.7
370	481.0	482.3	483.6	484.9	486.2	487.5	488.8	490.1	491.4	492.7
380	494.0	495.3	496.6	497.9	499.2	500.5	501.8	503.1	504.4	505.7
390	507.0	508.3	509.6	510.9	512.2	513.5	514.8	516.1	517.4	518.7
400	520.0	521.3	522.6	523.9	525.2	526.5	527.8	529.1	530.4	531.7

TABLE B-STANDARD TO ESTIMATE (Au): NO. 2 COLORS

NO. 3 COLORS

Weight in Mgrs.

Colors	0	1	2	3	4	5	6	7	8	9
		0.3	0.7	1.0	1.3	1.7	2.0	2.3	2.7	3.0
10	3.3	3.7	4.0	4.3	4.7	5.0	5.3	5.7	6.0	6.3
20	6.7	7.0	7.3	7.7	8.0	8.3	8.7	9.0	9.3	9.7
30	10.0	10.3	10.7	11.0	11.3	11.7	12.0	12.3	12.7	13.0
40	13.3	13.7	14.0	14.3	14.7	15.0	15.3	15.7	16.0	16.3
50	16.7	17.0	17.3	17.7	18.0	18.3	18.7	19.0	19.3	19.7
60	20.0	20.3	20.7	21.0	21.3	21.7	22.0	22.3	22.7	23.0
70	23.3	23.7	24.0	24.3	24.7	25.0	25.3	25.7	26.0	26.3
80	26.7	27.0	27.3	27.7	28.0	28.3	28.7	29.0	29.3	29.7
90	30.0	30.3	30.7	31.0	31.3	31.7	32.0	32.3	32.7	33.0
100	33.3	33.7	34.0	34.3	34.7	35.0	35.3	35.7	36.0	36.3
110	36.7	37.0	37.3	37.7	38.0	38.3	38.7	39.0	39.3	39.7
120	40.0	40.3	40.7	41.0	41.3	41.7	42.0	42.3	42.7	43.0
130	43.3	43.7	44.0	44.3	44.7	45.0	45.3	45.7	46.0	46.3
140	46.7	47.0	47.3	47.7	48.0	48.3	48.7	49.0	49.3	49.7
150	50.0	50.3	50.7	51.0	51.3	51.7	52.0	52.3	52.7	52.0
160	53.3	53.7	54.0	54.3	54.7	55.0	55.3	55.7	56.0	56.3
170	56.7	57.0	57.3	57.7	58.0	58.3	58.7	59.0	59.3	59.7
180	60.0	60.3	60.7	61.0	61.3	61.7	62.0	62.3	62.7	63.0
190	63.3	63.7	64.0	64.3	64.7	65.0	65.3	65.7	66.0	66.3
200	66.7	67.0	67.3	67.7	68.0	68.3	68.7	69.0	69.3	69.7
210	70.0	70.3	70.7	71.0	71.3	71.7	72.0	72.3	72.7	72.0
220	73.3	73.7	74.0	74.3	74.7	75.0	75.3	75.7	76.0	76.3
230	76.7	77.0	77.3	77.7	78.0	78.3	78.7	79.0	79.3	79.7
240	80.0	80.3	80.7	81.0	81.3	81.7	82.0	82.3	82.7	83.0
250	83.3	83.7	84.0	84.3	84.7	85.0	85.3	85.7	86.0	86.3
260	86.7	87.0	87.3	87.7	88.0	88.3	88.7	89.0	89.3	89.7
270	90.0	90.3	90.7	91.0	91.3	91.7	92.0	92.3	92.7	93.0
280	93.3	93.7	94.0	94.3	94.7	95.0	95.3	95.7	96.0	96.3
290	96.7	97.0	97.3	97.7	98.0	98.3	98.7	99.0	99.3	99.7
300	100.0	100.3	100.7	101.0	101.3	101.7	102.0	102.3	102.7	103.0
310	103.3	103.7	104.0	104.3	104.7	105.0	105.3	105.7	106.0	106.3
320	106.7	107.0	107.3	107.7	108.0	108.3	108.7	109.0	109.3	109.7
330	110.0	110.3	110.7	111.0	111.3	111.7	112.0	112.3	112.7	113.0
340	113.3	113.7	114.0	114.3	114.7	115.0	115.3	115.7	116.0	116.3
350	116.7	117.0	117.3	117.7	118.0	118.3	118.7	119.0	119.3	119.7
360	120.0	120.3	120.7	121.0	121.3	121.7	122.0	122.3	122.7	123.0
370	123.3	123.7	124.0	124.3	124.7	125.0	125.3	125.7	126.0	126.3
380	126.7	127.0	127.3	127.7	128.0	128.3	128.7	129.0	129.3	129.7
390	130.0	130.3	130.7	131.0	131.3	131.7	132.0	132.3	132.7	133.0
400	133.3	133.7	134.0	134.3	134.7	135.0	135.3	135.7	136.0	136.3

TABLE B-STANDARD TO ESTIMATE (Au): NO. 3 COLORS

NO. 4 COLORS

Weight in Mgrs.

Colors	0	1	2	3	4	5	6	7	8	9
				0.1	0.1	0.1	0.1	0.1	0.2	0.2
10	0.2	0.2	0.2	0.3	0.3	0.3	0.3	0.3	0.4	0.4
20	0.4	0.4	0.4	0.5	0.5	0.5	0.5	0.5	0.6	0.6
30	0.6	0.6	0.6	0.7	0.7	0.7	0.7	0.6	0.8	0.8
40	0.8	0.8	0.8	0.9	0.9	0.9	0.9	0.9	1.0	1.0
50	1.0	1.0	1.0	1.1	1.1	1.1	1.1	1.1	1.2	1.2
60	1.2	1.2	1.2	1.3	1.3	1.3	1.3	1.3	1.4	1.4
70	1.4	1.4	1.4	1.5	1.5	1.5	1.5	1.5	1.6	1.6
80	1.6	1.6	1.6	1.7	1.7	1.7	1.7	1.7	1.8	1.8
90	1.8	1.8	1.8	1.9	1.9	1.9	1.9	1.9	2.0	2.0
100	2.0	2.0	2.0	2.1	2.1	2.1	2.1	2.1	2.2	2.2
110	2.2	2.2	2.2	2.3	2.3	2.3	2.3	2.3	2.4	2.4
120	2.4	2.4	2.4	2.5	2.5	2.5	2.5	2.5	2.6	2.6
130	2.6	2.6	2.6	2.7	2.7	2.7	2.7	2.7	2.8	2.8
140	2.8	2.8	2.8	2.9	2.9	2.9	2.9	2.9	3.0	3.0
150	3.0	3.0	3.0	3.1	3.1	3.1	3.1	3.1	3.2	3.2
160	3.2	3.2	3.2	3.3	3.3	3.3	3.3	3.3	3.4	3.4
170	3.4	3.4	3.4	3.5	3.5	3.5	3.5	3.5	3.6	3.6
180	3.6	3.6	3.6	3.7	3.7	3.7	3.7	3.7	3.8	3.8
190	3.8	3.8	3.8	3.9	3.9	3.9	3.9	3.9	4.0	4.0
200	4.0	4.0	4.0	4.1	4.1	4.1	4.1	4.1	4.2	4.2
210	4.2	4.2	4.2	4.3	4.3	4.3	4.3	4.3	4.4	4.4
220	4.4	4.4	4.4	4.5	4.5	4.5	4.5	4.5	4.6	4.6
230	4.6	4.6	4.6	4.7	4.7	4.7	4.7	4.7	4.8	4.8
240	4.8	4.8	4.8	4.9	4.9	4.9	4.9	4.9	5.0	5.0
250	5.0	5.0	5.0	5.1	5.1	5.1	5.1	5.1	5.2	5.2
260	5.2	5.2	5.2	5.3	5.3	5.3	5.3	5.3	5.4	5.4
270	5.4	5.4	5.4	5.5	5.5	5.5	5.5	5.5	5.6	5.6
280	5.6	5.6	5.6	5.7	5.7	5.7	5.7	5.7	5.8	5.8
290	5.8	5.8	5.8	5.9	5.9	5.9	5.9	5.9	6.0	6.0
300	6.0	6.0	6.0	6.1	6.1	6.1	6.1	6.1	6.2	6.2
310	6.2	6.2	6.2	6.3	6.3	6.3	6.3	6.3	6.4	6.4
320	6.4	6.4	6.4	6.5	6.5	6.5	6.5	6.5	6.6	6.6
330	6.6	6.6	6.6	6.7	6.7	6.7	6.7	6.7	6.8	6.8
340	6.8	6.8	6.8	6.9	6.9	6.9	6.9	6.9	7.0	7.0
350	7.0	7.0	7.0	7.1	7.1	7.1	7.1	7.1	7.2	7.2
360	7.2	7.2	7.2	7.3	7.3	7.3	7.3	7.3	7.4	7.4
370	7.4	7.4	7.4	7.5	7.5	7.5	7.5	7.5	7.6	7.6
380	7.6	7.6	7.6	7.7	7.7	7.7	7.7	7.7	7.8	7.8
390	7.8	7.8	7.8	7.9	7.9	7.9	7.9	7.9	8.0	8.0
400	8.0	8.0	8.0	8.1	8.1	8.1	8.1	8.1	8.2	8.2

TABLE B-STANDARD TO ESTIMATE (Au): NO. 4 COLORS

TABLE C

LIST OF ABBREVIATIONS TO BE USED ON DRILL LOGS

Bld	Boulder	Wh	White
BR	Bedrock	Yl	Yellow
S	Sand	C	Coarse
Cal	Caliche	Dc	Decomposed
Cg	Conglomorate	F	Fine
Cl	Clay	Fl	Flaky
Gv	Gravel	Gn	Granular
Mud	Mud	H	Hard
Sh	Shale	L	Little
Sl	Soil	La	Loose
SS	Sandstone	M	Mostly
Tl	Tailings	Md	Medium
W	Water	Sft	Soft
Blk	Black	Sm	Some
Bl	Blue	Ti	Tight
Br	Brown	V	Very
P	Purple		Angular
R	Red		
Gr	Green		
Grey	Grey		

TABLE C-LIST OF ABBREVIATIONS TO BE USED ON DRILL LOGS

STANDARD VALUES FROM DRILL HOLE DATA

Size of Drive Pipe (in.)	Small 4	4	5	5	5	6
Drive Shoe - O.D. (in.)	5.25	5.25	6.125	6.50		7.50
Drive Shoe - I.D. (in.)	3.826	4.00	4.875	5.00		5.75
Drive Shoe - Mean Dia. (in.)	4.538	4.625	5.50	5.75		6.625
Eff. Dia. Shoe (in); based on 1.062 x Mean Dia.	4.82	4.91	5.84	6.10		7.04
Value of C = $\frac{4950.3}{D^2}$	212.0	204.7	145.0	133.0		100.0
Theor. Vol. (cy) x 1.05	0.00496	0.00514	0.00725	0.0079		0.0105
Casing - O.D.	4.50	4.75	5.5625	5.5625		6.625
Casing - I.D.	3.826	4.082	4.875	5.047		5.855
Theor. Rise (ft) x 1.05	1.68	1.52	1.51	1.53		1.52
Theor. Vol. (cy) (1 ft rise)	0.00296	0.00336	0.00480	0.00515		0.00693
Eff. Dia. Shoe (in); Mean Dia.	4.538	4.625	5.50	5.75		6.625
Value of C = $\frac{4950.3}{D^2}$	240.0	231.3	163.5	149.5		113.0
Theor. Vol. (cy) x 1.05	0.00438	0.00454	0.00642	0.00704		0.00928
Theor. Rise (ft) x 1.05	1.48	1.35	1.34	1.37		1.34
Theor. Vol. (cy); (1 ft Rise)	0.00296	0.00336	0.00480	0.00515		0.00693

TABLE D-STANDARD VALUES FROM DRILL HOLE DATA

Original Duplicate

Examination _Grey Pass_ .... Line................. Hole _116_ ...(2..) Sheet _1_ of _1_ Sheets

Elevation Collar............................ Co-ord. N.................., E................ Date Started _7-7-79_ ......194...

Offset from Stake, Bear.................;.........ft. hor.,........ft. vert.................. Date Compl. _7-11-79_ ......194...

FIELD LOG

Pump Time Hr. Min.	Depth Drilled Ft. 1/10	Drive	Core Before Pumping	After Pumping	Colors 1	2	3	4	Meas. Vol. Cu. Yds.	Correc-tions Mgs.	Est. Wt. Mgs.	Formation
9:15	10 0	10	0 6	8 0 3		1	··		0.046	—	0.7	Dug 2' grssn/
10:55	13 0	3 0	3 0	0 0 2					0.018	—		Gr & snd
11:50	17 0	4 0	4 0	4 5 0 3					0.030			
12:20	20 0	3 0	3 0	0 0 2					0.024		··	
1:00	23 0	3 0	3 0	9 1 5					0.030		··	
2:15	27 0	4 0	7 0	5 3 0					0.044			Rising
2:45	30 0	3 0	8 1	0 2					0.050			
8:15	32 0	2 0	3 7	2 5			··		0.020	—	0.2	Rising
8:30	34 0	2 0	4 9	1 3			···		0.028	-0.1	0.3	"
9:20	37 0	3 0	5 3	1 7		1	··		0.030	+0.1	0.7	"
10:00	40 0	3 0	5 7	0 7			···		0.035	-0.1	0.6	Gr sss
11:15	43 0	3 0	4 5	4 0					0.026			Rising
12:00	46 0	3 0	11 5	6 0			···		0:018			Rising
12:45	50 0	4 0	5 8	3 0					0.096		"	
2:10	53 0	3 0	10 4	3 0					0.060		"	
3:15	54 0	1 0	5 0	0 0		3	·		0.034	-1.2	1.6	
9:00	55 0	1 0	8 0	0		1	··		0.024	-0.2	0.7	
9:30	56 0	1 0	1 7	0 2		2			0.010	-0.1	1.0	
10:00	57 0	1 0	2 2	0 5					0.018			Rising
10:45	59 0	2 0	3 0	1	2	··		0.026	-1.1	4.7	Gr
11:25	60 0	1 0	1 7	0 0	1	4	···		0.014	-1.3	6.0	Gr
12:00	61 0	1 0	1 5	0 1		1			0.010	—	0.7	Gr ss
12:20	62 0	1 0	2 0	0 2		1	··		0.012	-0.2	0.7	
1:00	63 0	1 0	9 0	0		3	··		0.016	-0.7	2.1	
2:00	64 0	1 0	1 7	0 0		4 5	···		0.012	-1.3	7.1	
2:25	65 0	1 0	5 0	0	1	5 13	···		0.010	-0.9	13.5	
2:50	66 0	1 0	2 0	0 5	4	22 13	···		0.014	-8.9	41.3	
3:30	68 0	2 0	3 0	0 2	1	9 12			0.020	—	19.0	
9:00	69 0	1 0	1 8	0 5	1	2 9	···		0.018	-4.0	103	
9:25	70 0	1 0	2 0	0 1		3 7	···		0.016	-2.2	7.2	
9:40	71 0	1 0	1 5	0 1		1 1	···		0.008	—	1.8	
9:55	72 0	1 0	1 6	0 0					0.005		0.1	All in b.s.
				Slimes					0.028			

DRILL

Type & No. _Keystone_

Size Drive Pipe _5¾" I.D._

Dia. Drive Shoe _7¼" O.D._

TIME LOG

Moving _0.60 hr_

Drilling _20.80 hrs_

Pulling _3.60 hrs_

Delays..................

Total _25.00_

DEPTH, ETC.

Water Level...................

Overburden _0.0'_

Gravel _71.0'_

To Bedrock _71.0'_

Penetrated Bedrock _1.0'_

Total Drilled _72.0'_

Type Bedrock _Dec. ss (Old Man)_

CALCULATIONS

Calc. Vol. _0.72 cu. yd._

Meas. Vol. ....

Core Vol. ....

Drive Shoe Factor _.00'_

Core Factor _1.4_

Vol. Factor _2.01 cu. ft. drive_

Est. Wt. Mgs. _120.1_

Wt. Gold Mgs. _120.16_

Correction.....................

Corrected Gold, Mgs.............

Est. Fineness _930_

U. S. $ per Fine Oz............

Wt. Black Sand _Mix_

Working Depth................

Normal W. L...................

Dredging Depth Below, Normal W. L...........

PERSONNEL

Driller.....................

Foreman.....................

Calc. by.....................

Checked by..................

Engineer in Charge...........

* These entries must be completed in the Field. See over for Remarks.

Estimated Mean Value, U. S. cents per Cubic Yard

Pay Stratum.........ft. to ....ft = ...cents. (= ....mgs. per cu. yd.)

Tailings.........ft.   Virgin Ground.........ft.   3.340

Calculated Total Dredging Depth (excl. water).........ft. = ...cents. (= ....mgs. per cu. yd.)

LINE.........HOLE..........

SAMPLE DRILL LOG

Original Duplicate

Examination ...Grey Fly... ..... Line.................................. Hole 9C ...'9 (River) Sheet 1 of 1 Sheets

Elevation Collar...................... Co-ord. N....................., E............ Date Started June 20/79 194...

Offset from Stake, Bear.............;......ft. hor.,.......ft. vert............ Date Compl. June 22/79 194...

FIELD LOG

Pump Time		Depth Drilled		Core			Colors				Meas. Vol.	Correc- tions	Est. Wt.	Formation
Hr.	Min.	Ft.	1/10	Drive	Before Pumping	After Pumping	1	2	3	4	Cu. Yds.	Mgs.	Mgs.	
'/20 10:30	10	0	0	0	5	2	0	8			0.034		Nil	Dug 2', gr & snd
" :00	14	0	4	0	4	3	3	0			0.026		Nil	Rising gr(md) & snd
11:15	18	0	4	0	8	0	1	5			0.058		Nil	Gr(md) & snd
11:50	21	0	3	0	5	0	2	5			0.024		Nil	Gr(md)(w) & snd
12:05	25	0	4	0	7	3	0	8			::	0.050	-0.1	0.4	Rising smts
12:40	28	0	3	0	4	9	2	5			0.024	+0.2	0.2	" "
1:00	32	0	4	0	7	3	0	5			...	0.050	-0.1	0.3	" "
2:30	35	0	3	0	6	5	1	7	1	·	0.042	-0.1	0.5	" "
3:00	38	0	3	0	6	2	0	6			0.044		Nil	Gr & snd
4:15	40	0	2	0	4	0	0	5			0.024		Nil	" " "
'/21 8:10	42	0	2	0	5	7	0	5		·	0.036	-0.1	0.2	Loose gr & sn
9:10	45	0	3	0	5	0	1	0		·	0.040		0.1	" " "
9:40	48	0	3	0	6	7	6	0	?	::	0.094	—	Nil	Bld tr, na??m
10:05	50	0	2	0	3	2	2	4	1		0.092	-0.7	1.0	Rising gr & snd
11:15	51	0	1	0	10	7	0	3		1	4	::	0.070	-3.3	3.8	Boulder in sx
12:00	53	0	2	0	4	7	0	8		3	4	:::	0.026	-1.5	5.3	Hd dr, gr & sn (l cr)
12:15	54	0	1	0	2	6	0	0		5	3		0.022	-3.7	8.0	Gr & snd
12:50	55	0	1	0	4	0	0	0	1	8	45		0.030	-33.5	51.5	Nugget + 20 m in (19 c snt)
2:00	56	0	1	0	2	0	0	3		4	8		0.012	-1.7	9.5	Gr & snd
2:15	57	0	1	0	2	2	1	1		2	9		0.020	-1.5	7.5	" " "
2:30	58	0	1	0	3	1	1	5		4	17	:::	0.020	-6.8	14.5	Rising gr & snd
2:50	59	0	1	0	3	8	0	7		3	14	:::	0.022	-6.5	11.8	" " "
3:00	60	0	1	0	3	1	0	8			9	:::	0.026	-22	5.5	Gr & snd
3:35	61	0	1	0	2	5	0	5		3	::	0.010	-1.1	3.6	" " "
4:25	62	0	1	0	2	3	0	3		14	42	:::	0.012	-11.0	36.5	" " "
'/22 8:30	63	0	1	0	1	7	0	0		1	6		0.004	-0.8	4.5	" " some bk
9:20	63	5	0	5	1	2	0	0		1	2		0.062	-1.0	2.5	All in brk (Old Man)
									Slimes			0.024			
					1-46							0.679	-75.5	167.5	

DRILL *

Type & No. Keystone

Size Drive Pipe 5¾" ⌀

Dia. Drive Shoe 7/4" ⌀

TIME LOG *

Moving 1 day from Nelson Crk

Drilling 14:50

Pulling 1:50 (Pipe shrd)

Delays

Total 40:40 hrs

DEPTH, ETC. *

Water Level ≈ 2' below collar

Overburden 0'

Gravel 63'

To Bedrock 63'

Penetrated Bedrock .5'

Total Drilled 63.5'

Type Bedrock Old Man ss

CALCULATIONS

Calc. Vol. 0.635

Meas. Vol. 0.878

Core Vol. 1.415

Drive Shoe Factor .100'

Core Factor 1.4

Vol. Factor 0.01 / cu / feet dr

Est. Wt. Mgs. 167.2 mgs

Wt. Gold Mgs. 281.35 mgs

Correction 75.5 228/35 -12745 1672.5

Corrected Gold, Mgs. 154.30

Est. Fineness 930

U. S. ¢ per Fine Oz.

Wt. Black Sand Minimum

Working Depth

Normal W. L.

Dredging Depth Below, Normal W. L.

PERSONNEL

Driller * T. Kilkelley

Foreman * "

Calc. by * T.54

Checked by "

Engineer in Charge "

* These entries must be completed in the Field. See over for Remarks.

Estimated Mean Value, U. S. cents per Cubic Yard

Pay Stratum..........ft. to ....ft = ...cents. (= 242.99 mgs. per cu. yd.)

Tailings..........ft. Virgin Ground..........ft. 3750 grms/curyd

Calculated Total Dredging Depth (excl. water)..........ft. = ...cents. (= ....mgs. per cu. yd.)

LINE..............HOLE..........

9CK - '79 (River)

SAMPLE DRILL LOG

Original Duplicate

FIELD LOG

Examination .... _Green F..._ ... Line.................... Hole.... _/_ .... Sheet... _/_ ...of.....Sheets

Elevation Collar... _5.6.76..._ Co-ord. N.............., E................. Date Started... _.10.79_ **194**

Offset from Stake, Bear............;.......ft. hor.,........ft. vert............ Date Compl... _13.79_ **194**

Pump Time		Depth Drilled		Core			Colors				Meas. Vol. Cu. Yds.	Correc- tions Mgs.	Est. Wt. Mgs.	Formation
Hr.	Min.	Ft.	1/10	Drive	Before Pumping	After Pumping	1	2	3	4				
11:30	10	0		0	7	0	0	3				0.048	——	08	1 foot top...
12:00	14	0	4	0	4	3	1	0				0.034	——	0.5	Rising gra..
2:45	18	0	4	0	5	1	0	8				0.042	——		Grvl + sn..
1:30	21	0	3	0	4	8	2	0				0.036	——		Rising
2:25	25	0	4	0	7	0	8	2				0.044	——		C..p+sn..
		0	3	0	6	8	3	0				0.048	——		Grvl g+ 4s..
3:30	31	0	3	0		0	3	0				0.010	——		" " "
4:00	35	0	4	0	10	0	3	0				0.078	——		" " "
5:30	38	0	3	0	6	0	0	5				0.042	——		" " (fin)
9:35	41	0	3	0	5	1	1	2		1		0.036	——	0.7	" " "
10:50	44	0	3	0	1	5	3	2				0.038	——		" " "
11:25	47	0	3	0	8	8	1	2		5		0.054	-1.3	2.9	" " "
12:00	49	0	2	0	4	8	3	0		4		0.032	-0.8	2.3	Rising w..
12:45	52	0	3	0	7	3	0	7		1		0.020	-0.3	0.7	Md. g.. s..
1:10	55	0	3	0	5	8	3	0		1		0.046	+0.1	0.6	" " "
2:45	58	0	3	0	6	2	1	4		3		0.046	-1.6	2.2	S..cl..
4:10	60	0	2	0	5	1	2	2		2		0.032	-0.4	1.4	
8:50	62	0	2	0	5	2	8		12	2		0.020	+4.2	71.0	
9:15	64	0	2	0	4	5	1	3	7	2		0.024	-1.3	7.5	
9:50	65	0	1	0	5	0	0		10	20	58	0.016	-1.3	90.2	6" b..
11:00	65	5	0	5	0	8	0	0				0.004			
							Sli..		0.020						
													-63.0		

DRILL *

Type & No... _Keystone_
Size Drive Pipe... _5⁵⁄₄"ø_
Dia. Drive Shoe... _7 1/4 3_

TIME LOG *

Moving... _1 hr_
Drilling... _15 hrs_
Pulling... _3 hrs_
Delays... _..._
Total... _..._

DEPTH, ETC. *

Water Level at... _6 b.. in cellar_
Overburden... _1'_
Gravel... _63'_
To Bedrock... _64'_
Penetrated Bedrock... _1.5'_
Total Drilled... _65.2_
Type Bedrock... _(d.. ss)_

CALCULATIONS

Calc. Vol... _5.635_
Meas. Vol... _0.8.._
Core Vol... _..._
Drive Shoe Factor... _100'_
Core Factor... _1.4_
Vol. Factor... _1.4_
Est. Wt. Mgs. *... _253.8_
Wt. Gold Mgs... _370_
Correction... _-43.0_ (=.._
Corrected Gold, Mgs... _285.52_
Est. Fineness... _..._
U. S. $ per Fine Oz... _..._
Wt. Black Sand... _N..._
Working Depth... _..._
Normal W. L... _..._
Dredging Depth Below, Normal W. L... _..._

PERSONNEL

Driller *... _..._
~~Foreman~~ *... _..._
Calc. by... _..._
Checked by... _..._
Engineer in Charge... *
* These entries must be completed in the Field. See over for Remarks.

Estimated Mean Value, U. S. cents per Cubic **Yard** ___
Pay Stratum..........ft. to ....ft =..........cents. (= _35.98_ mgs. per cu. yd.)
Tailings..........ft.   Virgin Ground..........ft.   _6,727 o_
Calculated Total Dredging Depth (excl. water)..........ft. = ...cents. (= ....mgs. per cu. yd.)

SAMPLE DRILL LOG

Original Duplicate

I _o Consolidated Gold Dredg. g, Ltd.   **FIELD LOG**

Eng. From 20

Examination _7_____ Line _____ Hole _____ Sheet _1_ of _1_ Sheets
Elevation Collar ____ Co-ord. N_____, E _____ Date Started _____ 19__
Offset from Stake, Bear_____; ____ft hor., ____ft. vert. Date Compl._____ 19____

Pump Time	Depth Drilled	Core			Colors				Meas. Vol. Cu. Yds.	Correc- tions Mgs.	Est. Wt. Mgs.	Formation	DRILL *
Hr. Min	Ft. 1/10	Drive	BEFORE PUMPING	AFTER PUMPING	1	2	3	4					Type & No._____

DRILL *
Type & No._____
Size Drive Pipe_____
Dia Drive Shoe_____

TIME LOG *
Moving _5:50_
Drilling _13:5_
Pulling _____
Delays _____
Total _____

DEPTH, ETC. *
Water Level_____
Overburden _52.5_
Gravel _10.2_
To Bedrock _62.7_
Penetrated Bedrock _____
Total Drilled _____
Type Bedrock _____

CALCULATIONS
Calc. Vol._____
Meas. Vol._____
Core Vol._____
Drive Shoe Factor _100_
Core Factor _145_
Vol. Factor _75_
Est. Wt. Mgs. * _65.0_
Wt. Gold. Mgs. _150.3_
Correction $+\frac{37.2 \times 1503}{65.0} = +86.0$
Corrected Gold, Mgs. _236.3_
Est. Fineness _810_
U. S. $ per Fine Oz _35.00_
Wt. Black Sand Oz _4.8_
Bedrock elev. _180_
Normal W. L._____
Dredging Depth
Below Normal W. L. _64_

PERSONNEL
Driller * _____
Foreman * _____
Calc. by _H. Caballero R._
Checked by _H. Caballero R._
Engineer in Charge_____
* These entries must be completed in the Field.
See over for Remarks
B. R. Class No.

Estimated Mean Value: U. S. cents per Cubic Yard

Pay Stratum_____ft. to._____ft._____cents.
Tailings_____ft. Virgin Ground _62.7_ ft. Bedrock _0.3_ ft.
Calculated Total Dredging Depth (excl. water) _63.0_ ft. _35.9_ cents, (_375_ mgs. per cu yd.)
LINE _366_   HOLE _15c_

FORMA 94 - MARZO 69 500

Keystone   Tip. Libertad 7223-E

SAMPLE DRILL LOG

FIELD LOG

Examination Colombia Line I6 Hole # 140 Sheet 1 of 2 Sheets
Elevation Collar 266 Co-ord. N 97,436, E 91,877 Date Started 11:00 April 8 19 70
Offset from Stake, Bear ; 0.0' ft. hor., 0.0' ft. vert Date Compl 11:30 April 16 19 70

Pump Time Hr.	Depth Drilled Ft.	1/10	Core Drive	Core Before Pumping	Core After Pumping	Colors 1	Colors 2	Colors 3	Colors 4	Meas. Vol. Cu. Yds.	Corrections Mgs.	Est. Wt. Mgs.	Formation
4/8/70 2:30	Started Drilling												
3:00	5	-	5 -	1 7	- -	-	-	-	-	.005		—	yl. wh. f. md. c. fl. s. Ls.
3:45	10	-	5 -	3 3	- -	-	-	-	-	.017		—	" " " " " " "
4:15	15	-	5 -	4 7	- -	-	-	-	-	.030		—	" " " " " " "
4/13/70 8:20	20	-	5 -	5 0	- -	-	-	-	-	.038		-	" " " " " " "
8:45	25	-	5 -	7 1	2 0	-	-	-	5	.036		0.1	" " " " " " " f. md.
9:10	29	-	4 -	6 7	2 3	-	-	-	4	.032		0.1	Wh. grey. f. md. c. fl. s. f. Gu.
9:30	33	-	4 -	6 8	1 2	-	-	1	6	.034		0.4	" " " " " " " "
9:45	35	-	2 -	5 -	1 7	-	-	-	-	.019		-	" " " " " " " "
10:20	39	-	4 -	7 9	- -	-	-	-	4	.031		.031	" " " " " " " " yl. Fl
10:55	46	-	7 -	7 6	- -	-	-	-	-	.009			Grey. yl. Cl
11:55	50	-	4 -	3 5	- -	-	-	-	4	.011	+0.1	0.1	" " " f. md. c. Gu.
12:00	51	-	1 -	3-3	4 2	-	-	3	20	.014	(0)	1.4	Grey. wh. f. md. c. fl. s. f. md. c. Gu.
12:30	53	-	2 -	5 5	2 1	-	-	1	27	.010	(0)	0.8	" " " " " " " " "
12:45	55	-	2 -	5 0	2 -	-	-	-	3	.016		0.1	Ditto
12:50	56	-	1 -	3 2	1 2	-	-	-	1	.010		—	"
1:25	58	-	2 -	3 3	- 9	-	-	-	8	.014		0.2	"
1:35	59	-	1 -	2 4	- 7	-	-	2	40	.012	-0.2	1.5	"
1:40	60	-	1 -	2 -	- -	-	-	2	37	.007	-0.4	1.1	"
1:50	61	-	1 -	1 1	- -	-	-	2	20	.006	+0.4	1.1	"
[illegible]	62	-	1 -	1 7	- -	-	-	-	30	.008	-0.1	0.6	"
2:45	63	-	1 -	1 7	- 7	-	-	1	68	.011		1.7	"
3:00	64	-	1 -	1 8	- 5	-	-	6	34	.009	+0.3	2.7	"
3:20	65	-	1 -	1 3	- -	-	2	10	71	.006	+0.5	7.3	"
4:00	66	-	1 -	1 4	- -	-	1	17	237	.007	+0.4	11.7	"
4:15	67	-	1 -	1 4	- -	-	-	6	156	.007	+0.2	5.1	"
4:30	68	-	1 -	1 4	- -	-	-	8	158	.006	+0.2	5.9	"
4/17/70 8:00	69	-	1 -	1 4	- -	-	-	6	137	.006	+0.2	4.7	"
8:35	70	-	1 -	1 2	- -	-	-	7	133	.008	+1.0	5.0	"
9:15	71	-	1 -	1 4	- -	-	-	3	68	.007	+0.1	2.4	"
9:40	72	-	1 -	1 4	- -	-	1	3	70	.009	+0.1	3.7	"
10:00	73	-	1 -	1 3	- -	-	-	1	23	.006	+0.1	0.8	"
10:15	74	-	1 -	1 2	- -	-	-	1	23	.007	+0.2	0.8	"
10:50	75	-	1 -	1 4	- -	-	-	2	16	.007		1.0	"
11:10	76	-	1 -	1 4	- -	-	-	1	6	.007		0.4	"
11:55	77	-	1 -	1 1	- -	-	-	1	17	.003	+0.2	0.6	"

DRILL *
Type & No. Keystone #51
Size Drive Pipe 6"
Dia. Drive Shoe 7 3/8"

TIME LOG *
Moving
Drilling
Pulling
Delays
Total

DEPTH, ETC. *
Water Level
Overburden
Gravel
To Bedrock
Penetrated Bedrock
Total Drilled
Type Bedrock

CALCULATIONS
Calc. Vol
Meas. Vol
Core Vol
Drive Shoe Factor
Core Factor
Vol. Factor
Est. Wt. Mgs. *
Wt. Gold Mgs.
Correction
Corrected Gold, Mgs.
Est. Fineness
U. S. $ per Fine Oz.
Wt. Black Sand
Working Depth
Normal W. L.
Dredging Depth Below Normal W. L.

SEE Page #2

PERSONNEL
Driller *
Foreman *
Calc. by
Checked by
Engineer in Charge

* These entries must be completed in the Field. See over for Remarks.

Estimated Mean Value, U.S. cents per Cubic Yard Nil 20' • Only #3 ; #4 Colors taken for corrections
Pay Stratum ....ft. to ....ft = ...cents. (= ....mgs. per cu. yd.)
Tailings ....ft. Virgin Ground 89.4 ft. +0.6' B.R.
Calculated Total Dredging Depth (excl. water) 90 ft. = 18.1 cents. (= 189 mgs. per cu. yd.)
LINE I6 HOLE 140

SAMPLE DRILL LOG

FIELD LOG

Examination _Colombia_ .... Line _I 6_ .... Hole _# 140_ .... Sheet _2_ of _2_ Sheets
Elevation Collar ..... _266_ ..... Co-ord N. _97,936_ ..., E. _91,877_ ..... Date Started _Alice April 8_ 19 _70_
Offset from Stake, Bear .... ; ...... ft. hor., ... ft. vert .... Date Compl _11:30 April 16_ 19 _70_

Pump Time Hr.	Depth Drilled			Core			Colors				Meas. Vol. Cu. Yds.	Correc- tions Mgs.	Est. Wt. Mgs.	Formation
	Ft.	1/10	Drive	Before Pumping	After Pumping		1	2	3	4				
2:15	78	-	1	-	1 3	-	-	-	2	13	.006	+0.1	1.0	Grey wh. Rnd e Pl,s Rnd e Sa. H
1:45	79	-	1	-	1 5	-	-	-	1	10	.012	/	0.5	Ditto
2:00	80	-	1	-	2 7	-	7	-	-	1	7	.007	-0.1	0.4	"
2:20	81	-	1	-	2 4	-	-	-	2	6	.013	-0.3	0.8	"
2:30	82	-	1	-	1 8	-	-	-	3	12	.008	-0.2	1.2	"
2:35-83	83	-	1	-	1 8	-	-	-	4	3	.008	0.3	1.4	"
4:30	84	-	1	-	1 8	-	-	-	3	23	.011	-0.3	1.5	"
10:35	85	-	1	-	2 -	-	3	-	-	2	13	.006	-0.1	1.0	"
10:50	86	-	1	-	1 4	-	2	-	-	3	13	.008	+0.3	1.3	"
11:00	87	-	1	-	1 4	-	-	-	3	5	.008		1.1	"
1:10	88	-	1	-	1 6	-	-	-	-	3	.007		0.1	"
2:55	89	-	1	-	1 7	-	2	-	-	3	.008		0.1	"
1:20	90	-	1	-	1 9	-	-	-	-	3	.006		0.1	"
1:40	91	-	1	-	2 -	-	1	-	-	-	.003		-	M.R. Bl. Gn. B.R

Below Shoe

| 2:05 | 92 | - | 1 | - | - | - | 1 | - | - | - | .002 | | - | M.R. Bl. Gr. BR |
| 2:30 | 93 | - | 1 | - | - | - | 2 | - | - | - | .001 | | - | " " " " |

Repair

Total Colors | 4 | 08 | 53 | | 72.2

'13 Finished Drilling
11:30 Finished Pulling

Value per Milligram = 8 35 x .850 x 100 = .0956 ¢ / mg
_____
_31, 103.5_

Corrected Gold Mgs.   169.9 x 100 = 189.0 mgs./yd³
_____
_900_

_189.0 x .0956 = 18.1 ¢_

DRILL •
Type & No. _Keystone #51_
Size Drive Pipe .... _6"_
Dia. Drive Shoe .... _7 7/8 "_

TIME LOG •
Moving ..... _3:30_
Drilling ..... _21:15_
Pulling ..... _6:00_
Delays ..... _23:45_
Total ..... _54:30_

DEPTH, ETC. •
Water Level - _3.7' 11:30 4/16/70_
Overburden ..... _23.0_
Gravel ..... _66.4"_
To Bedrock - _89.4"_
Penetrated Bedrock _3.6"_
Total Drilled ..... _93'_
Type Bedrock _M.R. Bl. Gn. BR_

CALCULATIONS
Calc. Vol ..... –
Meas. Vol ..... –
Core Vol ..... –
Drive Shoe Factor ..... _100_
Core Factor ..... _1.95_
Vol. Factor ..... _7.5_
Est. Wt. Mgs. • ..... _72.2_
Wt. Gold Mgs ..... _162._
Correction +3.2 _42.7 = t 7.2_   _72.2_
Corrected Gold, Mgs. ..... _169.9_
Est. Fineness ..... _850_
U. S. $ per Fine Oz. _35.00_
Wt. Black Sand ..... _4.5_
Working Depth ..... _177_
Normal W. L. ..... _263_
Dredging Depth Below, Normal W. L. _87_

PERSONNEL
Driller • ....
Foreman • ....
Calc. by ....
Checked by ....
Engineer in Charge ....
• These entries must be completed in the Field. See over for Remarks.

stimated Mean Value, U. S. cents per Cubic Yard
Pay Stratum .... ft. to .... ft = ... cents. (= .... mgs. per cu. yd.)
Tailings .... ft.   Virgin Ground _89.4_ ft. _+0.6' BR_
Calculated Total Dredging Depth (excl. water) _90.0_ ft. = _18.1_ cents. (= _189_ mgs. per cu. yd.)
LINE _I 6_ .... HOLE _140_

SAMPLE DRILL LOG

FIELD LOG

Examination .... Line .... A .... Hole .... No. 2 .... Sheet .... 1 .. of .. 1 .. Sheets

Elevation Collar .... 1050' .... Co-ord. N ...., E .... Date Started .. Jan. 14 ... 19 81

Offset from Stake, Bear ....;.....ft. hor.,......ft. vert .... Date Compl. .. Jan 15 ... 19 81

Pump Time	Depth Drilled		Core			Colors				Meas. Vol.	Correc-tions	Est. Wt.	Formation
Hr.	Ft.	1/10	Drive	Before Pumping	After Pumping	1	2	3	4	Cu. Ft.	Mgs.	Mgs.	
+ 11:50			Started Drilling										
11:55	5	0	50			-	-	-	-	.110		—	Yl. Br. Clay + Sm. Gu
12:10	10	0	50			-	-	-	-	.275		—	" " " " Md. Gu
1:00	15	0	50			-	-	-	11	.440	-0.1	0.2	" " " " " "
2:05	20	0	50				1	7	55	.330	-1.1	4.3	" " " " " " "
2:15	25	0	50			2	4	10	86	.330	-1.1	16.2	" " " + F. + C. Gu.
2:22	30	0	50				7	13	71	.220	+1.8	15.2	" " " + " " "
2:30	35	0	50				1	4		.275		0.4	" " " + Sm. F. Gu.
3:02	40	0	50			-	-	-		.550		—	" " " No. Gu.
3:28	45	0	50			-	-	-		.500		—	Yl. Br. Clay
3:54	50	0	50			-	-	-				—	" " "
4:12	55	0	50			-	-	-				—	U. Hard @ 59' Bl. Clay
10:45	83	0	780	Rotary System								—	Compact org. H. Grey Bl. Cl
10:45				Finished Drilling									
11:45				Finished Pulling									
				Repair									
											-3.5	36.3	

Value per Milligram = $\frac{$480 \times .850 \times 100}{31,103.5}$ = 1.312 ¢

Corrected Gold Mgs. = $\frac{39.4 \times 550}{36.0'}$ = 526 mgs/yd³

526.0 × 1.312 = 690.0 ¢

DRILL • Rev. Cir
Type & No. Becker - R.C
Size Drive Pipe .. 5½" O.D.
Dia. Drive Shoe .. 3" cutting crowd o

TIME LOG •
Moving 1:00
Drilling 5:12
Pulling 1:00
Delays 1:20
Total 8:32

DEPTH, ETC •
Water Level .... -25.5'
Overburden .... 0.0'
Gravel .... 35.0'
To Bedrock W.B.R. 35.0'
Penetrated Bedrock .. 48.0'
Total Drilled .... 83.0'
Type Bedrock False. H. Bl. Gke

CALCULATIONS
Calc. Vol. .... —
Meas. Vol. .... —
Core Vol. .... —
Drive Shoe Factor .. 550
Core Factor Not in use
Vol. Factor .. 0.0491 ft³
Est. Wt. Mgs. • 36.3
Wt. Gold Mgs. = ... 38.1
Correction $\frac{-3.5 - 38.1}{36.3}$ = -3.7
Corrected Gold, Mgs. / 34.41
Est. Fineness 850
U. S. $ per Fine Oz. Est. 480
Wt. Black Sand .. 3.4 ozs.
Working Depth .. 36.0'
Normal W. L. .... —
Dredging Depth
Below, Normal W. L. .... —

PERSONNEL
Driller • ....
Foreman • ....
Calc. by ....
Checked by ....
Engineer in Charge ....
• These entries must be completed in the Field. See over for Remarks.

Estimated Mean Value, U. S. cents per Cubic Yard

Pay Stratum .... ft. to .... ft = ... cents. (= .... mgs. per cu. yd.)

Tailings .... ft.   Virgin Ground 35.0 ft. + 1' B.R.

Calculated Total Dredging Depth (excl. water) 36.0 ft. = 650.0 cents. (526 mgs. per cu. yd.)

LINE   A   HOLE No. 2

SAMPLE DRILL LOG

Form No. 1
Original—Duplicate

CONSOLIDATED
PLACER DREDGING, INC.
235 Montgomery St., Suite 2300
San Francisco, CA 94104

Sheet .... of .... Sheets

SHAFT LOG

COMPANY_____

Property................................ Area.. Line No.................... Shaft No....................

Elev. of Collar.......................Ft.   Co-ord. N.................... Co-ord. E....................

DEPTHS FROM SURFACE		VOLUME IN CU. FT.		NUMBER OF COLORS			GOLD IN MILLIGRAMS		FORMATION Gr=Gravel Sa=Sand Bo=Boulders B/R=Bed Rock c=Coarse f=Fine	MOISTURE Wet Moist Damp Dry	SHAFT SECTION	REMARKS
From Ft.	To Ft.	Bank Meas.	Box Meas.	No. 1	No. 2	No. 3	Est. Weight	Actual Weight			SURFACE	
TOTALS												

ESTIMATED MEAN VALUES PER CU. YD.

Pay Stratum ....Ft. to.................Ft. = ...Mgs. = ...Cents—Gold at   and..................Fine

Working Depth ....Ft. to.................Ft. = ...Mgs. = ...Cents— "   "   "   "   .... "

Total Depth ....Ft. to.................Ft. = ...Mgs. = ...Cents— "   "   "   "   .... "

SCREEN ANALYSIS	TIME	PERSONNEL
Oversize = + 6" = ...%	Date Started....................	Shaft Sunk by....................
Coarse = — 6" to + 1" =..................%	Date Completed....................	Sample Cut by....................
Medium = 1" to + ¼" = ...%	Hours Worked....................	Panned by....................
Fine = — ¼" = ...%	Delays....................	Weighed by....................
Black Sand by Bulk..................%	....	Calculations by....................
Size Shaft	....	Map Reference....................
New or Old Shaft....................	....	
Side Sampled	....	Checked and Approved,
....	....	....

SAMPLE SHAFT LOG

Form No. 1
Original—Duplicate

Sheet 1 of ...... Sheets

Sample
SHAFT LOG

COMPANY

[Prop]erty Terrace #1 Area Tropics Line No. — Shaft No. 1

Elev. of Collar ≈87 Ft. above river WL Co-ord. N. 5°25' Co-ord. E. 122°39'

[Da]te	DEPTHS FROM SURFACE		VOLUME IN CU. MT.		NUMBER OF COLORS			GOLD IN MILLIGRAMS		FORMATION Gr=Gravel Sa=Sand Bo=Boulders B/R=Bed Rock c=Coarse f=Fine	MOISTURE Wet Moist Damp Dry	SHAFT SECTION SURFACE	REMARKS
	From Ft.	To Ft.	Bank Meas.	Box Meas.	No. 1	No. 2	No. 3	Est. Weight	Actual Weight				
/12	0.0	1.0	1.0	2.0037					38.8		Wet		Black top soil w/heavy clay
/17	1.0	1.5	0.5	1.0588					3.0		"		Fine sand w/brown sticky clay
/19	1.5	2.0	0.5	0.8728					6.2		"		" " " " "
/20	2.0	2.5	0.5	0.9564					10.3		"		" " " " "
/22	2.5	3.0	0.5	0.8539					17.4		"		Brown coarse sand, gr. & clay
/25	3.0	3.5	0.5	0.8197					55.0		"		" " " " " "
/27	3.5	4.0	0.5	0.9791					451.0		"		Greenish brown coarse sand & gr. w/boulders ≈ 40 cm. dia., no clay
/?9	[illegible]	4.5	0.5	0.9905					756.0		"		" " " " " "
/?0	4.5	5.0	0.5	0.8197					300.0		"		" " " " " "
/?1	5.0	5.5	0.5	0.9563					27.9		"		Sand & gravel, cobbles ≈ 6cm ϕ
/?7	5.5	6.0	0.5	1.0247					142.7		"		" " " " " "
/?9	6.0	6.5	0.5	0.9108					21.9		"		Sand & gravel w/20cm thick lense of fine sand. No clay.
/?0	6.5	7.0	0.5	0.7742					58.9		"		Coarse sand & gravel w/60% 2cm ϕ pebbles
[T]OTALS													

ESTIMATED MEAN VALUES PER CU. YD.

Pay Stratum ....Ft. to ....Ft. = ...Mgs. = ...Cents—Gold at ...... and ....Fine
Working Depth ....Ft. to ....Ft. = ...Mgs. = ...Cents— " " " " .... "
Total Depth ....Ft. to ....Ft. = ...Mgs. = ...Cents— " " " " .... "

SCREEN ANALYSIS

Oversize = + 6" = ...%
Coarse = — 6" to + 1" = ...%
Medium = — 1" to + 1/4" = ...%
Fine — 1/4" = ...%
Black Sand by Bulk ....%
Size Shaft ....
New or Old Shaft ....
Side Sampled ....

TIME

Date Started ....
Date Completed ....
Hours Worked ....
Delays ....

PERSONNEL

Shaft Sunk by Hand
Sample Cut by ....
Panned by ....
Weighed by ....
Calculations by ....
Map Reference ....
Checked and Approved,

SAMPLE SHAFT LOG

Form No 1
Original—Duplicate

Sheet .... of ..... Sheets

SHAFT LOG
SAMPLE

COMPANY_____

roperty.................... Area.................... Line No.................... Shaft No....................

Elev. of Collar.................... Ft. .... Co-ord. N.................... Co-ord. E....................

DEPTHS FROM SURFACE		VOLUME IN CU. FT./MT.		NUMBER OF COLORS			GOLD IN MILLIGRAMS		FORMATION Gr=Gravel Sa=Sand Bo=Boulders B/R=Bed Rock c=Coarse f=Fine	MOISTURE Wet Moist Damp Dry	SHAFT SECTION SURFACE		REMARKS
From Ft.	To Ft.	Bank Meas	Box Meas.	No. 1	No. 2	No. 3	Est. Weight	Actual Weight					
7/13	7.0	7.5	0.5	0.6192					31.0		Wet			20 cm ⌀ boulders Gravel ∮ sand, No clay
7/14	7.5	8.0	0.5	0.7052					41.5		"			Coarse sand ∮ gravel w/30cm thick lens of fine sand
7/15	8.0	8.5	0.5	0.6536					197.9		"			Coarse sand ∮ gravel w/10% 10cm∮-20cm∮ boulders by rol
7/16	8.5	9.0	0.5	0.6880					223.6		"			" " " " " " "
7/18	9.0	9.5	0.5	0.7568					137.6		"			" " " " " " "
7/20	9.5	10.0	0.5	0.6020					99.7		"			" " " " " " "
7/21	10.0	10.5	0.5	0.6708					306.9		"			Coarse sand ∮ gravel 40 cm ⌀ boulders
7/23	2.5	11.0	0.5	1.0148					303.9		"			" " " "
7/27	11.0	11.5	0.5	0.5332					372.8		"			Coarse gravel ∮ sand ± 40% -10-40 cm ⌀ boulders
7/30	11.5	12.0	0.5	0.9138					602.8		"			No clay Gravel ∮ sand w/boulders
7/30	12.0	12.4	0.4	0.5160					117.9		"			Bedrock - conglomerate
TOTALS			12.4	20.6944					4324.7					Direct Eval = $\frac{4324.7}{20.6944}$ = 208.98 c

ESTIMATED MEAN VALUES PER CU. YD.

Pay Stratum ....Ft. to....................Ft. = ...Mgs. = ...Cents—Gold at .... and....................Fine

Working Depth ....Ft. to....................Ft. = ...Mgs. = ...Cents— " " " .... "

Total Depth ....Ft. to....................Ft. = ...Mgs. = ...Cents— " " " .... "

SCREEN ANALYSIS	TIME	PERSONNEL
Oversize = + 6" = ...%	Date Started....................	Shaft Sunk by....................
Coarse = — 6" to + 1" =....................%	Date Completed....................	Sample Cut by....................
Medium = — 1" to + ¼" = ...%	Hours Worked....................	Panned by....................
Fine = — ¼" = ...%	Delays....................	Weighed by....................
Black Sand by Bulk....................%	....	Calculations by....................
Size Shaft	....	Map Reference....................
New or Old Shaft	....	
Side Sampled	....	Checked and Approved....................

SAMPLE SHAFT LOG

TABLE 1 TENOR COMPARISONS

	ZONE A gr/cy	K/B	ZONE B gr/cy	K/B	ZONE A & B gr/cy	K/B	3:1 SITES	K/B
BECKER	0.71	1.51	0.61	1.35	0.61	1.48	1.60	
"(new factors)	1.04	1.02	0.86	0.96	0.88	1.02	1.09	
KEYSTONE	1.07		0.82		0.90			
BECKER HOLES	44		44		94		36	
KEYSTONE HOLES	21		14		36		12	
AREA (acres)	111		99		231			

KEYSTONE BECKER RATIOS

28 KEYSTONE HOLES, 81 BECKER HOLES	KEYSTONE	BECKER	RATIO
Actual Wt/Measured Volume	1.32 gr/cy	0.75 gr/cy	1.76:1
(corr'd wt X core factor) Divided by depth	1.00 gr/cy	0.66 gr/cy	1.51:1

ORE RESERVE ESTIMATE-ZONE A

#1:#2 21 Keystone, 44 Becker Holes	1.07 gr/cy	0.71 gr/cy	1.51:1
#1:#3 Becker calc. @ 3.125" core 38% expansion	1.07 gr/cy	1.03 gr/cy	1.04:1
#1:#4 Becker calc. @ 2.70" core 33% expansion	1.07 gr/cy	1.02 gr/cy	1.05:1

LINEAR REGRESSION, 3:1 Holes, Grey River

Becker calculated 3.125" core, no expansion	1.60:1
Becker calculated 3.125" core, 38% expansion	1.08:1

TABLE 2 - CORE RISE MEASUREMENTS

A. BECKER

	6 HOLES	4 HOLES
Advance	74 ft	60 ft
Measured	54.5 ft	44 ft
Core Rise	42.5 ft	34.7 ft
% of Theoretical	78%	79%
Measured Volume		3.215 ft^3
% of Theoretical		101%

B. KEYSTONE

7 HOLES		
Advance	176.5 ft	Core Pumped Out
Measured	176.5 ft	Measured Volume
Core Rise	220.4 ft	% of Theoretical
% of Theoretical	89%	5.75" I.D. Casing
Measured Volume	51.57 ft^3	
% of Theoretical	98%	

C. BECKER WITH KEYSTONE PIPE

	TEST 1	TEST 2
Drive	24"	37"
Core Rise	19.5"	28.75"
% Theoretical	58%	55%
Core Retained in Pipe	19.5"	28.75"
Measured Volume	0.424 ft^3	0.605 ft^3

BECKER TEST
TABLE 1-TENOR COMPARISONS
TABLE 2-CORE RISE MEASUREMENT

CORE VOLUME COMPARISONS - BECKER DRILL   (Volumes expressed in cubic feet.)

HOLE # 1

Drive	Meas. Vol.	Theor. Vol.	% of Theor.
0 - 10'	.660	.490	135
10 - 15'	.330	.245	135
15 - 20'	.330	.245	135
20 - 25'	.330	.245	135
25 - 30'	.385	.245	157
30 - 35'	.275	.245	112
35 - 40'	.220	.245	90
40 - 45'	.385	.245	157
45 - 50'	.220	.245	90
50 - 55'	.275	.245	112
	3.410	2.695	126

HOLE # 2

Drive	Meas. Vol.	Theor. Vol.	% of Theor.
0 - 5'	.110	.245	45
5 - 10'	.275	.245	112
10 - 15'	.440	.245	180
15 - 20'	.330	.245	135
20 - 25'	.330	.245	135
25 - 30'	.220	.245	90
30 - 35'	.275	.245	112
	1.980	1.715	115

HOLE # 3

Drive	Meas. Vol.	Theor. Vol.	% of Theor.
0 - 5'	.220	.245	90
5 - 10'	.385	.245	157
10 - 15'	.440	.245	180
15 - 20'	.330	.245	135
20 - 25'	.385	.245	157
25 - 30'	.330	.245	135
30 - 35'	.385	.245	157
35 - 40'	.330	.245	135
40 - 45'	.770	.245	314
45 - 50'	.605	.245	247
50 - 55'	.385	.245	157
55 - 60'	.385	.245	157
60 - 65'	.385	.245	157
65 - 70'	.385	.245	157
70 - 75'	.770	.245	314
75 - 77'	.165	.098	168
77 - 80'*	.165	.147	112
80 - 85'	.165	.245	67
85 - 90'	.110	.245	45
	7.095	4.410	161

HOLE # 4

Drive	Meas. Vol.	Theor. Vol.	% of Theor.
0 - 5'	.330	.825	40
5 - 10'	.110	.825	13
10 - 15'	.385	.825	47
15 - 20'	.220	.825	27
20 - 25'	.165	.825	20
25 - 30'	.165	.825	20
30 - 35'	.220	.825	27
35 - 40'	.165	.825	20
	1.760	6.600	27

* Rotary unit was used from 77 to 90'.

HOLE # 5

Drive	Meas. Vol.	Theor. Vol.	% of Theor.
0 - 5'	.220	.245	90
5 - 10'	.385	.245	157
10 - 15'	.220	.245	90
15 - 20'	.275	.245	112
20 - 25'	.330	.245	135
25 - 30'	.275	.245	112
30 - 35'	.330	.245	135
	2.035	1.715	119

HOLE # 6

Drive	Meas. Vol.	Theor. Vol.	% of Theor.
0 - 5'	.220	.245	90
5 - 10'	.275	.245	112
10 - 15'	.275	.245	112
15 - 20'	.220	.245	90
20 - 25'	.330	.245	135
25 - 30'	.165	.245	67
30 - 35'	.165	.245	67
35 - 40'	.275	.245	112
	1.925	1.960	98

Hole # 7

Drive	Meas. Vol.	Theor. Vol.	% of Theor.
0 - 5'	.220	.245	90
5 - 10'	.275	.245	112
10 - 15'	.220	.245	90
15 - 20'	.275	.245	112
20 - 25'	.330	.245	135
25 - 30'	.330	.245	135
30 - 35'	.275	.245	112
35 - 40'	.220	.245	90
40 - 43'	.330	.147	224
	2.475	2.107	117

	Average % of Theor. Vol.
Hole # 1	126
Hole # 2	115
Hole # 3	161
Hole # 5	119
Hole # 6	98
Hole # 7	117
Total ave. % (less hole # 4)	123

TABLE 3-CORE VOLUME COMPARISONS-BECKER DRILL (HOLES 1-7)

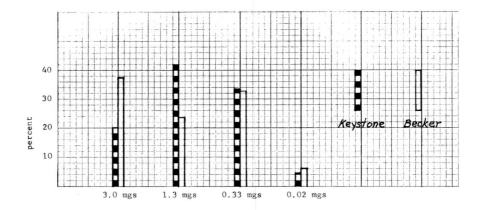

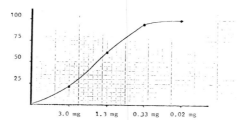

FIGURE 2.   Cumulative Weight Percent, Keystone Drill

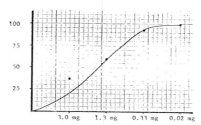

FIGURE 3.   Cumulative Weight Percent, Becker Drill

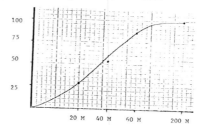

FIGURE 4.   Cumulative Weight Percent, Screened Sizes - Taramakau River.

FIGURES 1-4: DISTRIBUTION OF GOLD BY WEIGHT IN FOUR SIZE RANGES

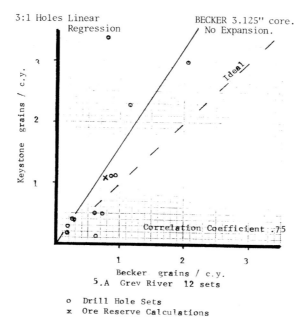

5.A Grey River 12 sets

o Drill Hole Sets
x Ore Reserve Calculations

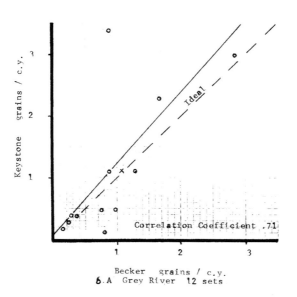

6.A Grey River 12 sets

FIGURES 5A AND 6A-LINEAR REGRESSION

FIGURE 5

3:1 Holes
     Linear
     Regression

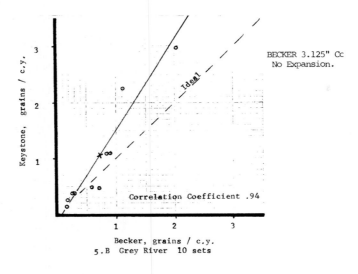

BECKER 3.125" Cc
No Expansion.

FIGURE 6.   3 : 1   HOLES LINEAR REGRESSION, Becker

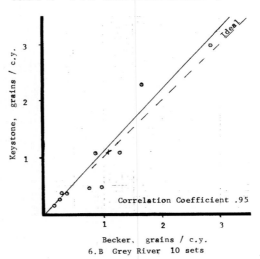

FIGURES 5B- AND 6B-LINEAR REGRESSION

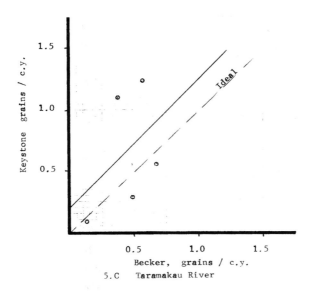

5.C   Taramakau River

3.125" core, 38% Expansion. .

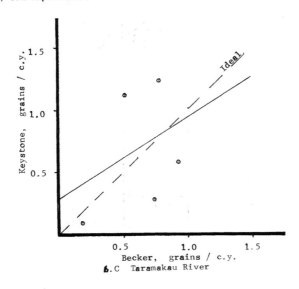

6.C   Taramakau River

FIGURES 5C- AND 6C-LINEAR REGRESSION

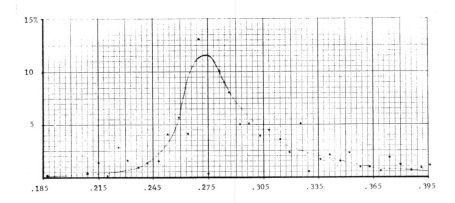

FIGURE 7.A.   Percent Distribution Of Measured Volumes, (excluding volumes greater than
0.4 cubic feet per 5 foot drive)   79 Becker Holes Studied

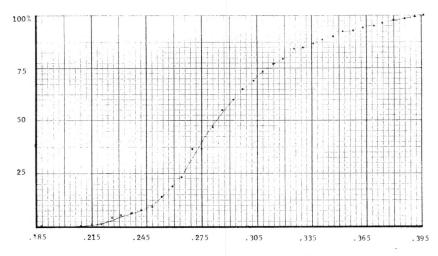

FIGURE 7.B.   Cumulative Percent Distribution of above Measured Volumes.

FIGURES 7A- AND 7B-DISTRIBUTION OF MEASURED VOLUMES

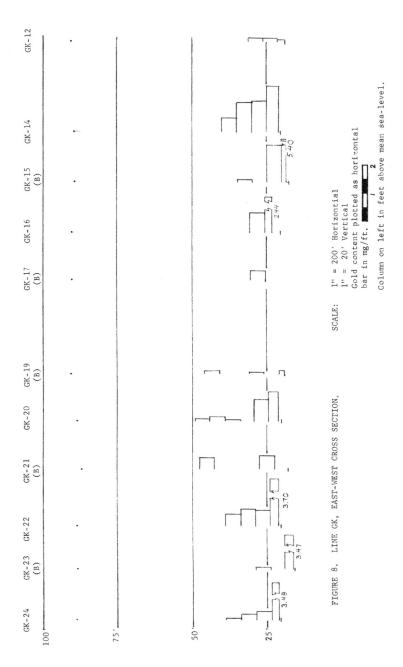

FIGURE 8. LINE GK, EAST-WEST CROSS SECTION.

SCALE: 1" = 200' Horizontal
        1" = 20' Vertical
        Gold content plotted as horizontal
        bar in mg/ft.

        Column on left in feet above mean sea-level.

FIGURE 8-LINE GK, EAST/WEST CROSS-SECTION

CHAPTER II
FEASIBILITY ANALYSIS-
PLACER MINING DECISIONS

Section 1
Principle Elements For Analysis

Once the prospect evaluation has been completed in a preliminary or final phase, a feasibility study should be conducted. The principal elements of this study should be as follows:

1. Tenor of the target mineral in place (mg/yd^3).

2. Volume of material that must be excavated and processed to produce the estimated quantity of target mineral, applying the average tenor from #1 above (i.e. Troy oz Au/yd^3).

3. The character of the soil and environment that must be mined and processed.

4. The equipment configuration to be used that is most suitable to the above conditions.

5. The financial limitations or strength of the investors to decide the magnitude of the operation that will govern the choice to be made in equipment and planning to proceed.

6. The projected operation and costs related thereto based upon the equipment decisions made, combined with the projected price of the target mineral thus providing the essential dollar values to be used in the cash flow analysis.

Equipment Tradeoff Analysis

The following examples will use gold as the target mineral and proven mining systems for analysis of choices. Figure 1-Placer Gold Mining/ Equipment Tradeoff, shows a division between a portable washing plant and a bucket ladder mining dredge (BL/M), that is primarily involved with the excavation rate of the equipment. A fixed price of $300./Troy ounce of gold, is used on the abscissa and computed against monthly production to produce revenue.

The ordinate is the monthly production rate in cubic yards of the dredge or a washing plant, being fed by equipment such as backhoe, dragline or frontend loader. The three trend lines are for different tenor levels of the deposit; i.e. 150, 350 and 500 mg/yd^3.

An operating cost line is described on the left based upon upgrading of dredging equipment as the volume increases. Assumptions of the size of the dredges in terms of bucket volume (ft^3), are delineated on the right margin. The dredge size selection is based upon efficient dredging operations and equipment as experienced over some 60 years of CPD's operations with gold mining using BL/M dredges. The profitability can be measured horizontally at any point on the graph, from the operating cost line on the left, right to the slope that is appropriate to the tenor of the deposit.

Production Cutoffs

The assumptions as to the equipment size selection is expanded upon in Figure 2 Cost of Operation Trends With Production Levels. The cutoff line for the maximum, practical production rate of a washing plant has been selected as 100,000 yd^3/month, and assumes an average production level of 200 yd^3/hr when operating, combined with an efficiency factor of 70% (i.e. actual bucket fill and operating time vs. theoretical 100% fill and operating time).

Our experience suggests that production above 100,000 yd³/month, introduces problems of multiple handling of material in the case of a portable plant. It is therefore more efficient to upgrade the equipment to a floating dredge at that point, if possible.

There are also such factors as the excavation device and its ability to clean bedrock. This is usually limited to a backhoe when dealing with portable equipment, combined with the feeding capacity of the plant to deal with ever increasing volumes of material in a single discharge.

There is a difficulty with a dragline for that reason, since the plant can only operate efficiently when there is some uniformity of material flow through the system. The tendancy with a dragline is to increase the volume of the bucket while attempting to maximize production, but this compounds the problem of dump volume.

Not considered on Figures 1 or 2, is the total gold-bearing volume of material that is proven in the deposit. If the tenor is, for example 500 mg/yd³ but only 300,000 yd³ total volume, the decision as to capital investment in the mining equipment would have to be balanced between minimizing the duration of the mining and the capital cost of the equipment.

A 14 ft³ BL/M dredge, for example would mine out the above deposit in one month. However, it would take about 1.5 years to mobilize a dredge of that size along with a total capital cost exceeding $5.0 million. A washing plant, on the other hand in that low volume would complete the mining in a few months. Add to that the fact that the plant could be mobilized in less than six months, for less than $1.0 million. This also points out the fact that the example of total volume rules out any commercial sized mine in any event.

It is also important to note that the depth of the deposit is not considered but again, the increased rate of dredging is assumed to be combined with a larger volume of material and a greater depth of deposit. When the gold is lying near bedrock and the tenor at that level is high, it must be diluted when calculating the full amount of material to be removed.

Thus a lower tenor may be acceptable which requires a larger dredge and higher volume production in order to move the material in a reasonable period of time and at a lower cost-per-unit volume. Also, as the rate of dredging increases it is expected that the cost-per-cubic yard will decrease. If it doesn't then something is wrong with the design of the equipment and/or the operation that should be corrected.

In this section, I have attempted to make a case for a cutoff of 100,000 yd^3/month for the production level of a portable washing plant. The suggestion is that above that level a dredge should be substituted. This means that the plant and equipment must be operating on a three-shift, seven days-per-week basis. Should that level of production not prove feasible, then production would be reduced accordingly.

Figure 3 Reserve Criteria vs. Tenor, further dramatizes a practical cutoff of tenor for a small mine, this time using $400/oz. and a fixed production rate of 100,000 yd^3/month.

It is important to note in Figure 2, the trend of diminishing returns as the plant production decreases, with the lines of tenor beginning to converge and profitability turning to loss. Likewise, however, the consequences of HIGH production in a portable plant as far as the material handling problem is concerned, are too often overlooked in the design stage. A high production level is essential to profitability and a cash flow analysis should demonstrate that fact but material handling problems must be in balance with other considerations.

While it might be possible to find a high grade deposit in small canyons, for instance, it is usually difficult to adequately sample such a deposit to determine the average tenor and whether it is possible to maintain the necessary production. As the cost of operation line clearly shows, the cost drops markedly once you shift to a dredge and further as production volume and size of dredge increases. The cost of operation of a small plant therefore must be kept under control and rate of production optimized within practical limits of the equipment and the deposit.

From Figures 1 & 2, we can arrive at the conclusion that depending upon tenor, as production drops below 100,000 yd³/month, profit degrades and a production level of 50,000 yd³/month may be considered as near the minimum level for a commercial-sized operation. To this should be added the constraint that the tenor must at that minimum, be no lower than 400 mg/yd³ combined with a floor of $300./Troy oz. It should be recognized though that in placers, an average tenor that high is rare.

With the above criteria established, we can arrive at a conclusion about the volume of the deposit. Unless there aresome environmental or other restrictions that would prevent the use of a BL/M dredge, it would be safe to say that a total volume of 8-10 million cubic yards would be the maximum for a portable plant. There would be grey areas to be considered that would establish a band of variability, but each case would finally be decided on its own set of conditions.

SECTION 2
EQUIPMENT SELECTION

In this section, I have excluded consideration of all other types of dredges in deference to a BL/M dredge and this needs some qualification. The number of suction type dredges, for instance, that have been installed in gold placer operations over the years probably could not be counted, they are so numerous. However, we do not know of a single suction dredge operation that has produced a profitable gold placer mine.

The exception to this is where a company or financial group controls a number of small suction dredges, using low-cost labor. Their costs are low and they profit off the sweat of those operators and divers. We recommend against this type of mining but that does not deter the thousands of small miners in developing countries nor their syndicators.

This is not intended to denegrate suction dredges in general since they have performed a valuable function in non-mining, navigational dredging (i.e., cutter, trailing hopper and plain suction). For industrial mineral mining such as sand and gravel, rutile, phosphates, the CS/M and in the last several years, the bucketwheel suction dredge (BWS/M), have proven to be effective systems.

The fact remains that the various types of suction dredges have their appeal when considering capital costs which are usually less than the BL/M dredge. This in turn helps to draw in the investors who would like to keep their investment and "estimated risk" at a minimum and can be easily persuaded that the technical capability is at least equal between dredge types.

It is interesting to note that in a publication dated circa 1938, concerning placer mining with portable units, that the same problem existed then. History has shown that in the vast majority of cases, the only commercially successful dredges mining gold, have been BL/M. A suction dredge regardless of its excavating device, carries too much water, does not clean bedrock and disperses fine gold at the suction mouth (except for BWS/M). These problems are serious detriments to a profitable operation in placer gold deposits.

The Illustration Section (Figures P10, P12-13, P18, P23, P25.1, P26), shows large scale, BL/M gold mining dredges producing between 300,000 and 500,000 yd³/month. Also shown is a floating washing plant in Brazil (Figure P11), fed by backhoe and using Yuba jigs for recovery of tin.

Portable Wash Plants

In selecting the configuration of equipment for small scale mining which we have now assumed would be in the range of 50-100,000 cuyd/month, we can assume that the excavation equipment, whether backhoe or other configuration must be able to clean bedrock and efficiently feed the hopper of the washing plant. We will therefore at this point look at the vital components of that plant (Figure P11).

Hopper Feed

The hopper must be of sufficient size to accept the volume of material that is discharged by the loader and not choke up as it disperses the material into the trommel. There must also be some water fed to the hopper to lubricate the surface but at the same time, the opening must be sufficiently constrictive that it doesn't permit too much material to enter the trommel at one time.

A set of grizzly bars are placed over the hopper discharge chute, to feed the trommel. This will absorb some of the shock loading of cobbles and boulders. A spacing of eight inches or more should be used so that values will not be lost. Where the name of the game is "saving gold," this can spell the difference between profit and loss.

Trommel Screen/Classifier

The proven system for classification in placer mining is the revolving trommel screen. The advantage of this equipment is that cobbles and small boulders can be discharged into it and thus be washed of clay and other material clinging to the rock that often contains gold, or, other target mineral.

This is the problem for instance, with shaking screens which are often used by mining operators in the interest of handling excess water from suction dredges and avoiding the higher cost of a stoutly built trommel. Since a flat screen cannot normally sustain large rocks or boulders falling on its surface, they narrow down the grizzly spacing and thus lose a portion of the gold.

A revolving trommel with screen plates in the periphery is therefore the recommended approach. The tread rings, structure, or ribbing, the drive mechanism and thrust bearings must all be built of sufficiently heavy material that they will last through the total period of the project under severe loading (see CHAPTER III MINING SYSTEMS FOR PLACERS, for details on trommels).

With the foregoing considerations, the decision can be made as to the diameter and length of the trommel along with the types of screen plates, size and shape of holes, to achieve a flow rate that is necessary to handle the material. The slope of the screen in its axis, is another point of decision. The size of screen openings also depends upon the type of primary mineral jig or other concentrating device used and its capability to handle larger particle sizes.

Critical Questions

When making decisions about placer gold mining equipment for small scale projects and attempting to evaluate the various products that are being promoted, the following questions need to be asked:

1. Can it process an average volume of over 50,000 yd³/month without significant interruption; (i.e. over 70% operating time).

2. Will the disposal of oversize and sluice discharge be handled adequately and provide for reclamation, with a minimum of multiple handling of material.

3. Is there an amalgamation system included that will not discharge mercury into the environment that provides continuous gold recovery without interrupting production.

4. Can the system operate for $1.00-1.50/yd³, including excavation equipment, overhead and G & A expense.

Table 1 is an example of an economic analysis for feasibility decision, of a small scale, gold placer and is based upon some of the foregoing factors.

TABLE 1
ECONOMIC ANALYSIS OF A GOLD PLACER,
PORTABLE WASHING PLANT OPERATION

ASSUMPTIONS
1. Average tenor of deposit 250mg/cuyd
2. Total proven reserves 8,000,000 cuyd's
3. Average production ratde 100,000 cuyd/month
4. Total period of mining 6.7 vears
5. Total investment with interest $1,500,000.
6. Cost of operation $1.00/cuyd
7. Fixed price of gold $300/Troy ounce

REVENUE   MONTHLY OPERATION
Production X Tenor X Price of Gold:
100,000 x 0.250/31.1 X 300. = $241,000.

OPERATING COST
100,000 x 1.00 = $100,000.

NET OPERATING PROFIT = $141,000.
TOTAL NET PROFIT
6.7 x 12 x 141,000. = $11,336,400.
Less capital investmen = $1,500,000.
Net Before Royalty or taxes = $9,836,400.

OPERATING COST ELEMENTS INCLUDE:
Labor & engineering
Fuel & maintenance
Utilities
Equipment rental
Insurance
Personnel benefits
Administration
Replacement spares

SECTION 3
BL/M DREDGES, LARGE SCALE PLACER MINING

In approaching the decision process for large scale placer mining using BL/M dredges, there are several principal factors to analyse. CHAPTER III MINING SYSTEMS FOR PLACERS, is intended to provide the most detail for this purpose while this Chapter emphasises some of the key elements influencing decisions.

Figure 2-Cost of Operation Trends, provides an indication of how cost of operation can vary with size of dredge in terms of bucket volume, along with approximate production level. Figure 4 BL/M Dredges Production Experience, gives some examples of actual production; dramatizing the point of diminishing returns on size increases.

In Table 2, BL/M Dredge Optimization, examples of actual dredges and their operation are indicated for comparison. An arbitrary Factor "f" is derived by dividing estimated capital cost of each dredge by the average production. The lower the figure, the higher the effective rate of return on investment. It is seen that the 14-ft^3 BL/M dredge carries the most favorable ratio.

As an additional comparison on a gross basis of assumptions, Figure 5 BL/M Dredges, Capital Cost vs. Production, demonstrates how capital cost tends radically upward after passing a middle point (near to a 14 ft^3 dredge). This gives some validity therefore to a point of size optimization, where multiple dredges vs a single large one, can be considered.

Dredge Mobilization
When the decision of dredge size for a given proven reserve is made, the capital cost budgeted including infrastructure and logistics support requirements considered, the time factor to complete mobilization must be planned. Figure 6 Mobilization Milestones, is an example of the principal elements of mobilizing a BL/M dredge on

site for mining ending with commencement of production. This represents a 7 ft³ BL/M dredge and would vary depending on size and digging depth as well as other factors. Some of the subsystems are not broken down including processing plant with mineral jigs but are included in time factors.

In Figure 7 Cash Drawdown Schedule, the same dredge funding requirements are shown. This provides an important financial planning guide for the investors, project manager, purchasing agent and accounting staff of the builder.

The fact that all funding needs to be committed and available to the project manager should be obvious; to prevent any lag in placing orders with sufficient lead time, taking advantage of early pay discounts, negotiating best prices and meeting shipping schedules as well as personnel manning. Budget overruns can easily be created by such delays and other problems that may relate to local conditions of assembly, weather, lease commitments, can be aggravated as well.

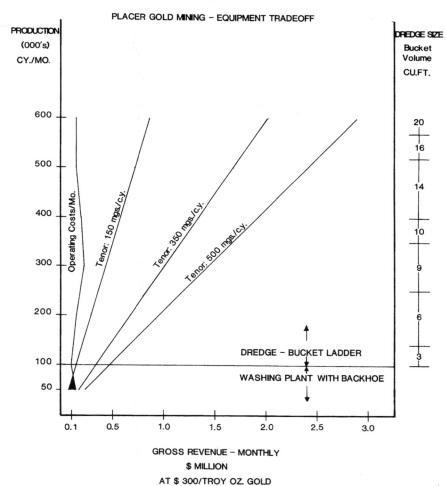

FIGURE 1

PLACER GOLD MINING – EQUIPMENT TRADEOFF

FIGURE 1-PLACER GOLD MINING, EQUIPMENT TRADEOFF

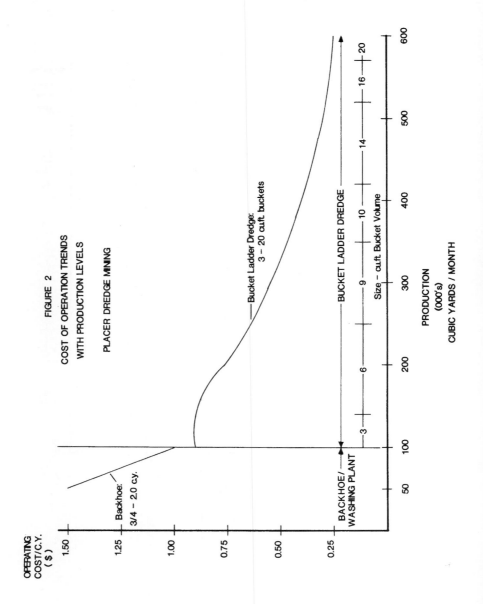

**FIGURE 2-COST OF OPERATION TRENDS WITH
PRODUCTION LEVELS, PLACER DREDGE MINING**

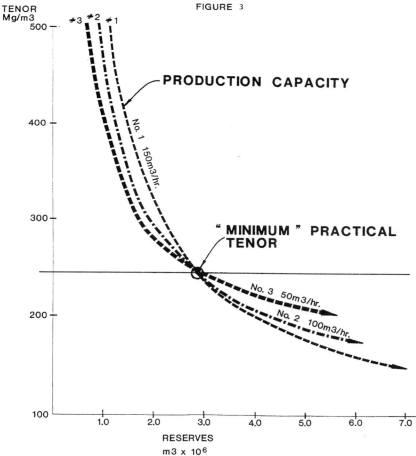

FIGURE 3-RESERVE CRITERIA VS. TENOR, SMALL PLACER GOLD MINES

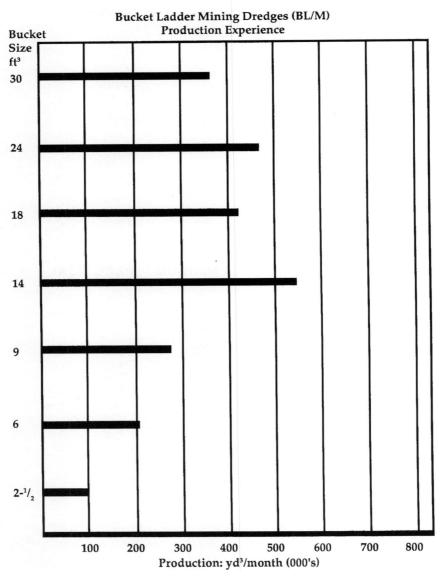

FIGURE 4-BUCKET LADDER MINING DREDGES (BL/M) PRODUCTION EXPERIENCE

TABLE 2

BL/M DREDGE OPTIMIZATION

BUCKET SIZE, COST & PRODUCTION
RELATIONSHIPS

BUCKET FT³	DEPTH		PRODUCTION(000)		CAPITAL COST US$-MIL.	FACTOR f
	m	ft	m³/mo	yd³/mo		
2-1/2	10.7	35	76.3	100	1.5	15.0
6	15.3	50	160.3	210	3.5	16.7
9	21.4	70	213.8	280	4.5	16.1
14	30.5	100	420.0	550	6.5	11.8
18	38.1	125	328.3	430	15.0	34.9
24	45.8	150	358.8	470	25.0	53.2
30	50.0	164	274.8	360	40.0	111.1

FOOTNOTES:

1.   Examples of size and production are based upon known BL/M dredges.

2.   Capital costs are either updated estimates or published actuals.

3.   Factor "f" = capital cost/yd³ of production, showing a relationship of cost effectiveness; i.e., the smaller the number the more efficient the operation.

TABLE 2-BL/M DREDGE OPTIMIZATION

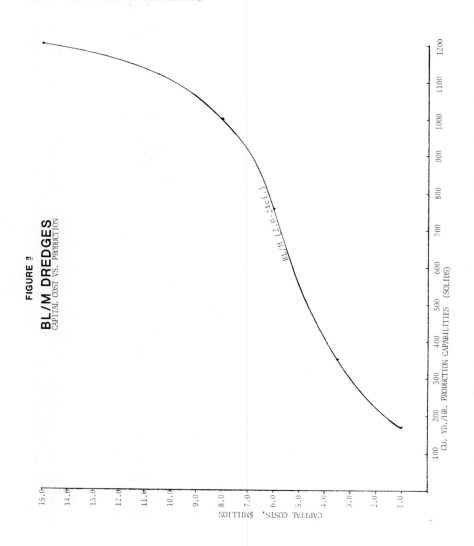

FIGURE 5-BL/M DREDGES: CAPITAL COST VS. PRODUCTION

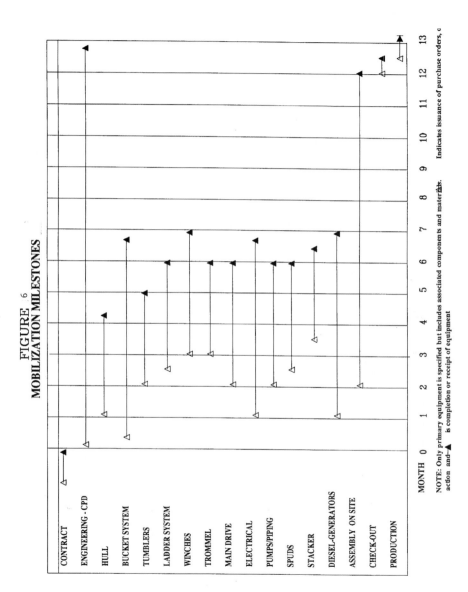

FIGURE 6-MOBILIZATION MILESTONES

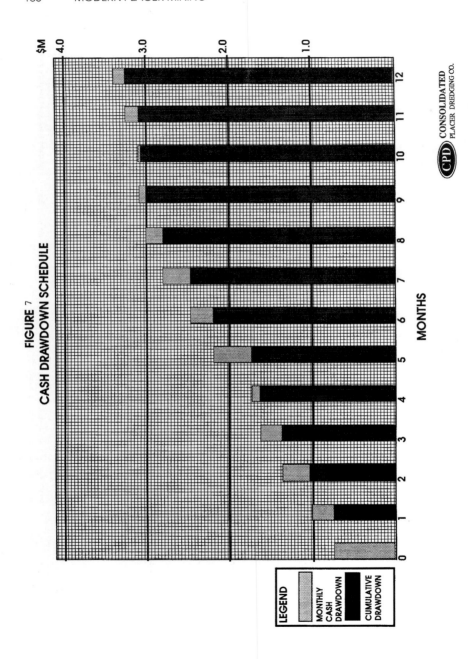

FIGURE 7
CASH DRAWDOWN SCHEDULE

CPD CONSOLIDATED PLACER DREDGING CO.

FIGURE 7-CASH DRAWDOWN SCHEDULE

ILLUSTRATIONS

FIGURES P1 WARD HAND DRILL:
P1.1LAND OPERATIONS/P1.2RIVER OPERATIONS WITH RAFT TETHERED

FIGURE P2 EMPIRE DRILL AND CREW, BRAZIL

FIGURE P3.0 "AIRPLANE" CHURN DRILL,
DRILLING TO 60 FT IN JUNGLE RIVER FLOOD PLAIN, PHILIPPINES

FIGURE P4.0 "KEYSTONE-TYPE"
CHURN DRILL, SAMPLING LARGE GOLD PLACER, GREY RIVER, NEW ZEALAND

FIGURE P5.0 BECKER HAMMER,
REVERSE CIRCULATION DRILL SAMPLING TO 110 FT., GREY RIVER, NEW ZEALAND

FIGURE P6.0
BULK SAMPLING BY SHAFTING, PERUVIAN ANDES AT
17,000 FT. ELEVATION FEEDING TROMMEL FROM SHAFT DIGGING-
SCREENED UNDERSIZE THROUGH "LONG TOM," WORKERS BREAKING UP CLAY

FIGURE P7.0
BADE CAISSON DRILL, 28" DIA, BULK SAMPLING PLACER DIAMOND DEPOSIT,BRAZIL

FIGURE P8.0 BADE CAISSON DRILL

FIGURE P9.0 YOST CLAM DRILL, CAISSON BULK SAMPLING

FIGURE P10-GOLD PLACER MINING DREDGES:
P10.1 BL/M, 14 FT³, RIO NECHI, COLOMBIA; P10.2 BL/M, 9 FT³, NOME, ALASKA

FIGURE P11.0 FLOATING WASH PLANTS, TIN PLACER FED BY BACKHOE; BRAZIL

FIGURE P12 GOLD PLACER MINING DREDGES
P12.1 BL/M DREDGE-10.5 FT³, BULOLO, NEW GUINEA; (P12.2) BL/M-11 FT³BOLIVIA

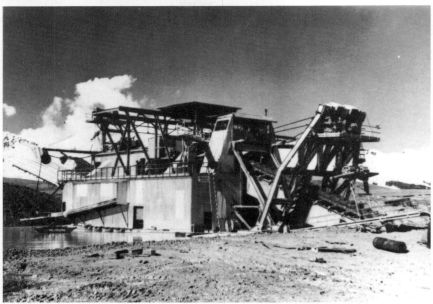

FIGURE P13.0 GOLD PLACER DREDGES; P13.1 BL/M, PATO 7A,"
14.5FT³, RIO NECHI, COLOMBIA; 13.2 BL/M 9 FT³, PAMPA BLANCA, PERU

FIGURE P14.0 TIN PLACER DREDGE;
P14.1 BL/M, 24 FT³, MALAYSIA; P14.2 BL/M TIN DREDGE 14 FT³, BOLIVIA

**FIGURE P15.0 MINING DREDGES; P15.1 BL/M TIN DREDGE, 22FT³,
INDONESIA; P15.2 S/M SAND & GRAVEL, AT 250 FT DEPTH, HOLLAND**

***FIGURE P16.0 RUTILE MINING DREDGES; P16.1 CS/M WITH
PROCESSING PLANT, AUSTRALIA; P16.2 CS/M 16," FLORIDA, USA***

FIGURE P17.0 TH/M SAND & GRAVEL MINING DREDGES, ENGLISH CHANNEL

*FIGURE P18.0 BL/M GOLD MINING DREDGE, 20 FT³, MINING
200 FT. BELOW SURFACE; YUBA PLACER GOLD CO., MARYSVILLE, CALIF., USA*

FIGURE P19.0
BWS/M TIN PLACER MINING DREDGE, WITH FLOATING PROCESSING PLANT, BRAZIL

FIGURE P20.0 DREDGE MINING; P20.1 BWS/M RUTILE DREDGE WITH PROCESSING PLANT, AUSTRALIA; P20.2 CS/M SAND & GRAVEL DREDGE FOR DEEP DIGGING

FIGURE P21.0 DS/M DREDGE MINING SALTS, DEAD SEA, JORDAN

FIGURE P22.0 S/M DREDGES, GOLD PLACER MINING, SMALL-SCALE OPERATIONS

FIGURE P23.0
TWO BL/M GOLD DREDGES, 14 FT³, RIO NECHI , ANTIOQUIA, COLOMBIA

FIGURE P24.0
BL/M DIAMOND PLACER MINING, 12 FT³ DREDGE, MINAS GERAIS, BRAZIL

FIGURE P25.0
PLACER MINING DREDGES; P25.1 BL/M 20 FT³, GOLD DREDGE TARAMAKAU
RIVER, SO ISLAND, NEW ZEALAND; P25.2 BL/M TIN DREDGE, 6 FT³, BOLIVIA

FIGURE P26.0
GOLD PLACER DREDGES; P26.1 BL/M, 18 FT³, 124 FT DEPTH,YUBA RIVER,
CALIF., USA; P26.2 BL/M GOLD DREDGE, 6-FT³, DUNKWA GOLD FIELDS, GHANA

FIGURE P27.0 BL/M TIN DREDGE, 12 FT³, "EURO -TYPE," BURMA

FIGURE P28.0 "SUNGEI PANDAN" BL/M TIN DREDGE, 9 FT³, MALAYSIA

FIGURE P29.0 BL/M TIN DREDGE, 24 FT³, SELANGOR #2, INDONESIA/BANKA ISLAND

FIGURE P30.0 BL/M TIN DREDGE, 24 FT³, "PERANGSANG #2," MALAYSIA

FIGURE P31.0 "BELITUNG #1, "BL/M TIN DREDGE, 22 FT³, INDONESIA

FIGURE P32.0 DREDGE MINING "PIONEERS," NEWTON CLEAVELAND (L),
W.P. HAMMON (R), C.1908, ON BL/M GOLD DREDGE, CALIFORNIA, CIRCA 1908

CHAPTER III
MINING SYSTEMS FOR PLACERS

The foregoing chapters have outlined the methods of finding and evaluating placer deposits with an emphasis on gold. Reference has been made to various types of mining equipment in CHAPTER II FEASIBILITY ANALYSIS. This chapter will look at the principal BL/ M dredge mining subsystems that are appropriate for placers.

Section 1
Categories of Placer Mining Equipment

Any responsible engineer given the job of selecting appropriate equipment for efficient mining of a given deposit, must balance a desire to apply new technology to proven systems with the need for an economic return. If there is no known way to mine a deposit and therefore something altogether new must be designed, the choice may be somewhere in between existing hardware and state of the art.

In the case of placer mining, there is considerable history combined with operating data that can be referred to in applying existing designs to mine a new deposit. If after making all calculations a more economical and dependable system can be found, then the improved methods should be used. But building of new, full-scale design "prototypes" for first time operation in the field without prior testing, is not considered good engineering nor responsible to investors.

If new designs are compelling, they should be thoroughly tested in small scale prototypes before making decisions on full scale construction. Improvements in technology including computer controls and automation, materials and components, should be evaluated for substitution throughout in proven systems. Parametric studies of alternative mining systems should be made to determine the optimum efficiency and economic tradeoffs in order to make rational, engineering decisions.

Given those basic assumptions, the following are the dredge mining systems generally available that may be considered for placer deposits of various minerals. (Note: Use of category codes from *WORLD DREDGING Mining & Construction magazine*, Directory of Worldwide Dredge Fleets. The symbol "M" is added to the dredge category symbols to indicate a mining application).

Dredge Mining Equipment
Bucket Ladder - BL/M
Bucket Backhoe - BB/M
Bucket Dragline - BD/M
Bucket Clamshell/Grab - BC/G/M
Bucketwheel Suction - BWS/M
Cutter Suction - CS/M
Plain Suction - S/M
Trailing Suction Hopper - TH/M
Dustpan Suction - DS/M

Section 2
BL/M Dredges for Placer Mining

The background of the BL/M dredge is well documented in "History of Mining Alluvial Gold" by Romanowitz (Appendix), outlining its early beginnings before mining, followed in the late 1800's when the system was assembled for mining in New Zealand. Since those early times, billions of cubic yards of material have been dredged and processed to recover gold; in California, Alaska, Canada, Colombia, New Guinea, Australia, New Zealand, China and Russia; to name the principal areas.

Gold dredging was overlapped by tin dredging beginning in 1922, in Malaysia, followed by Thailand, Burma, Indonesia, Tazmania, Brazil, and Nigeria with dredging volumes probably exceeding gold production. Most of this mining has been accomplished with BL/M dredges until recent times when backhoes and bucketwheel suction dredges have been used mainly in Brazil.

BL/M Types
Mining configurations of BL/M dredges originated in New Zealand. One of the design engineers emigrated from New Zealand to San Francisco, California, USA; another to London, England. Without belaboring the details of what evolved, the result was two different structural approaches; 1) the California type, and 2) the European type. Photographs are shown in the Illustration Figures and should be referred to with the following descriptions and discussions.

California-type BL/M Dredge
Dredges of this type built in years past which are still operating in some parts of the world today, include Yuba, Natomas, Bethlehem, Bucyrus-Erie, and Marion, as well as Simons-Lobnitz under license from Yuba. Distinctive features include an exposed bow gantry (i.e., no housing above it), winches mounted on the main deck, low center of gravity including upper tumbler and main drive motors and

gearing. This resulted in hull sizes being proportionately smaller than their European counterparts. The largest built was 18 ft³ buckets with maximum digging depth of 124 feet. Later retrofits changed to 20 ft³ buckets. Figure 1 is a profile drawing of a Yuba dredge designed for 18 ft³ buckets digging to 85 feet.

European-type BL/M Dredge

These designs originate from F.W. Payne & Sons, engineers in England. Most of the applications have been in tin deposits of Malaysia and Thailand. Dredges have been built in various shipyards under license or direction of Payne, but others have included IHC Holland (and its predecessor companies in Holland); Orenstein& Koppel, Germany; Simons-Lobnitz/Alluvial Dredges Ltd, Scotland, various builders in Europe and Japan.

Their distinctive has been an extended housing over the bow, mounting the winches high in the superstructure, and raising the upper tumbler and associated drive and gearing two- to three-times the height of the Cal-type. The largest mining application was the "BIMA" dredge, with 30 ft³ buckets to dig 160 feet below waterline, upper-tumbler 75 feet above deck and 10 feet freeboard.

DESIGN COMMENTS OF F.W. PAYNE & SON LTD., UK.
by J.A. Hewitt, Director

"Generally our dredges have been 'custom' designed to suit individual applications. In all cases, because of our ever increasing knowledge and experience with on-going bucket dredging operations, new dredges incorporate modifications and innovations to achieve greater operational and maintenance capabilities and also take advantage of updated technology being constantly applied to proprietary equipment and plant. From one dredge to the next these changes may not appear significant but over a period of ten plus years and say ten plus dredges they can be seen, collectively to be most significant.

One example of change to reduce maintenance difficulty and downtime was to replace the 'caterpillar' system of supporting long underslung bucket chains with a three independent idler system. Another, to reduce operational downtime, was to replace a conventional drop chute system incorporating an overburden sliding scuttle with a rotating head which when in appropriate radial positions, can feed bucket discharge to two screens simultaneously, to either one of the two screens or by-passes it directly to an overburden chute. The rotating head is power driven. This change means 10/15 minutes of downtime is saved every time the dredge changes to and from overburden stripping.

There have been many such changes which have contributed to higher efficiencies or ease of operation and maintenance. Over the years new technology involving Hydraulics, Pneumatics, Electrics, Automation etc., have been introduced or updated as appropriate. However it must be appreciated that in some locations which are very remote and/or the quality of labour and service is low, the dredges have to be designed as simply as possible with the minimum requirement for sophisticated operation, maintenance and service, albeit at the expense of efficiency. (For illustrations of Payne designed dredges, see Figures P15.1, P27-31).

Production Experience of BL/M Dredges
Table 1 BL/M Dredge Production, provides a comparison in selecting a placer mining system to show experience using BL/M dredges in a variety of conditions; from tropics to arctic regions. The list is not complete but draws upon prior studies that illustrate their capabilities in contrast with other systems. The company or dredge name, size of buckets, and highest monthly production are stated.

It should be noted that the listed production figures represent a variety of soil conditions; from sand and small gravel, to cobbles, boulders, cemented gravels, caliche, and permafrost sand, gravel and peat, and heavy clays. However, they nevertheless demonstrate the capabilities of BL/M dredges to produce on a 24-hour/day, 365 day/year schedule.

Bucketline Speed
Tests were conducted on some of the Colombian dredges of Pato Consolidated Gold Dredging Ltd., and International Mining Corp.(merged in 1962), to increase bucketline speed above standard levels. While wear on pins, buckets, tumblers and rollers, was increased the net result was positive, significantly increasing production.

This then became the standard approach for PATO. For instance, the dramatic production level of the 14 ft^3 BL/M Pato dredges up to 550,000 cuyd/month, was made possible by increasing average bucket speed from 22-24, to 30-32 bpm.

Other leading companies such as Yuba, were reluctant to follow the recommendations of Pato/CPD after studies of their operations to increase bucketline speed. Hence, the 18 ft^3 BL/M dredges' production did not reach the levels of the 14 ft^3 dredges. Some factors in support of Yuba's position, however, included the difference in material being dredged in California versus Colombia; i.e., heavy gravels and boulders vs clay, sand and gravel. Also, it was stated later by Romanowitz that the larger dredges such as 18 ft^3, required increased time for routine maintenance and repair because of size and weight of components, that dampened otherwise increased production for the bucket size.

The above experience can be contrasted with dredges in Malaysia, mining in soft clays and sands. Some of the 24 ft³ BL/M dredges digging to 50 meters, have had production closer to 900,000 yd³/month though not reported as an average. The "BIMA" 30 ft³ BL/M dredge when mining in offshore alluvials near Banka Island, Indonesia, was reported to average about 350,000 m³/month.

This can be considered an illustration of a design that passed the point of diminishing returns, since the theoretical production capability is about 1.0 million m³/month. In Malaysia, some of the 24 ft³ dredges will run the bucketline at 34 bpm when passing through soft clay. However, the higher-speed requires an active program of bucket and pin welding and changing along with lubrication of wearing surfaces and bearings.

BL/M Dredge Subsystems

Table 2 BL/M Dredge Segregations, represents a standard segregation of main components or subsystems, of a BL/M dredge. The numbers have been used with Cal-type dredges for many years and provide a ready cross-reference to costs, weights and specifications that are valuable when performing feasibility and parametric studies.

Table 2 provides points of design and application engineering for technology transfer opportunities. Some of these aspects and choices will be discussed in the following sections.

BL/M Dredge Size Selection

There are a number of factors in the selection of dredge size that should be considered in the Feasibility Study of a placer project. The total volume of material to be dredged, the dimensions of the deposit in area, depth and character of material to be dredged, are considerations that should be entered into the equation. Normally, the cost of extraction will be reduced as the size increases. However, this must be coupled with other factors such as the above. This is discussed in some detail in CHAPTER II FEASIBILITY ANALYSIS.

An example of size impact on production can be seen from Table 3 BL/M Dredge Size vs. Production Capabilities. For ease of comparison, an efficiency level of 65% (composite of average bucket fill and operating time vs. theoretical), and average bucket speed of 30 bpm, are used. It should be noted, that 65% was the average efficiency over many years of operation of both Yuba in California and Pato in Colombia.

Capital Cost of BL/M Dredges
The design features of BL/M dredges will influence the cost of the completed system. Factors such as bucket size and digging depth will be the largest determinates, producing an expanding cost of much of the structure; i.e., hull, superstructure, main drive, stacker. Capital costs of the above sizes assuming that a 6 ft^3 dredge would be designed to dig 40 ft. at one extreme, to a 24 ft^3 dredge digging to 150 ft.; might vary for a Cal-type from $3.0 to $20 million. This would also depend on where the assembly of the dredge took place, the source of the materials, and other factors.

Hull Design and Construction
The size of the hull is generally governed by the digging depth and bucket size of the dredge, followed by the area on deck necessary to accomodate the processing plant and other equipment. When an estimate of the above factors is determined, buoyancy, trim, stability calculations and metacentric height can be completed.

Sufficient freeboard must be calculated into the configuration to provide for safety and insurability. Operating conditions, river restrictions and ease of mobility are other elements that must be considered in optimizing the final hull dimensions.

Examples of hull dimensions on actual operating dredges, corresponding to bucket size and digging depth, are shown in Table 4. Note however, that these examples do not necessarily suggest that they are the most appropriate or adequate.

Too often hulls are designed for minimum freeboard and as new equipment is added over the subsequent years of operation, the freeboard decreases. This creates a sensitive stability and flotation condition where inattention to slow leaks and flooding of compartments, can result in capsizing. Therefore, a margin of safety at the initial or reconstruction phase, can prove to be the cheapest insurance available.

There was a trend at one time to construct hulls in multiple pontoons, to facilitate moving the dredge to other properties. While this increased the amount of steel and fabrication costs, it paid for itself in ease of a move. In recent years there has been a tendancy to abandon the old hull when moving the dredge to new locations, building a new one on site and having it ready when the superstructure, digging and processing plant arrive. For remote regions where transportation and labor skills are limited, the use of the pontoons still has decided advantages for speed of mobilization and ultimate cost.

An integral hull, fabricated and compartmentalized as with a barge or other vessel, has several advantages in its initial construction. The ordering of precut steel plate and its relative ease of shipment, is a large factor in such a cost study. This is true when setting up operations in third world countries, where transportation, roads and facilities are limited. In some cases, complete dredges broken down into manageable units, have been flown into remote areas such as New Guinea, Colombia, Guyana and Bolivia complete with pontoon sections.

Designing and building hulls to rules of recognized Societies (i.e. Lloyds Register, American Bureau of Shipping, Bureau Veritas), helps in obtaining insurance at favorable rates. When planned for inland use such as ponds, it is not as essential. But when operating in navigable waterways, rivers, harbors or offshore, it is important that such certifications be obtained.

Digging System
Bucket Design. An evolution of bucket designs occured throughout this century, based upon the material being dredged combined with wear. It was determined in the late 1800's that manganese added to steel, produced the most durable casting. The characteristic of manganese in steel resulted in an increase in durability with use and made it ideal for BL/M dredges. The same alloy or similar was applied to other heavy wear materials on the dredge, including upper and lower tumblers, heel plates, hopper discharge, pins and bushings, and ladder rollers.

A flared and tapered bucket was developed that improved cutting forces considering the manner of the swinging of the dredge to port and starboard. The gold dredging grounds of California and New Zealand influenced the rugged design of buckets, considering the large cobbles and boulders in those areas. For the most part, those designs were carried into other areas of the world with some variations.

For instance, CPD added such features as a "clay ring" inside the bucket to release suction caused by thick clay and help to ease discharge of material. Sources of managanese buckets have reduced considerably but foundaries with molds of some sizes and designs, may be found in Japan, Malaysia, Australia, Austria, UK and Brazil. There may be some possibilities still in Canada and the USA but would have to be carefully evaluated as to the experience of foundry personnel with large castings made with manganese; an art that is not easily passed on.

Digging Ladder

The length of the ladder in order to provide the design digging depth of the dredge, is determined by two basic factors. First is the digging angle to be used at the maximum depth. Cal-type dredges generally limited the depressed angle to 47 degrees for gold in heavy gravels. Malaysian dredges of Euro-type and digging in soft clay and jungle materials, depress their ladders to 60 degrees.

The other primary factor in length is determined by the selected height of the upper tumbler. The Euro-type dredge has tended to maintain the same geometry of the ladder-tumbler relationship. Thus when increasing digging depth the tumbler has been moved upward in proportion, placing emphasis upon balancing lateral forces on the hull. In comparison, the Cal-type dredge has placed an emphasis on keeping the center of gravity lower and not raising the upper-tumbler any more than necessary to maintain a gravity flow height for ease of feeding the treatment plant.

Another difference in the two dredge designs in relation to the ladder and upper tumbler height, is where the bow lines are attached. The Cal-type attaches them to a sheave on each side of the lower-end of the ladder. The Euro-type attach the lines to the bow deck on port and starboard, thus causing heavy load forces at the end of the ladder to be transmitted up the ladder, against the hull where contact is made and on to the upper tumbler. This aspect has been pointed out by Cleaveland in numerous of his papers and may have been adopted by some Euro-type dredge operators in the meantime.

Past recommendations have been made by Cal-type dredge designers, to lower the upper tumbler to 20 feet above deck, for a 190-foot digging depth, digging at 50 degrees. This creates problems in handling bucket discharge and pumping slurry through the treatment plant, but the alternatives have reasonable tradeoffs. The saving in structural size and cost of the dredge would be significant.

A structural girder design for ladders was evolved, coping with difficult digging conditions. For the most part, this has been the standard. Attempts to make structural analyses of these designs and loading conditions, in order to use higher tensile steels and less material have not met with alot of success.

The forces and dynamic loading on the ladder when digging hard materials on a continuous 24 hr day basis, are difficult to estimate and some shortcuts have resulted in structural failures. As an example of ladder mass, a 160-foot-long ladder will weigh approximately 2,000 lb/ft, or, total of 320,000 lbs. That is without the added weight of buckets, pins, rollers, tumblers and ore in the buckets. It is not uncommon for ladders to crack or break in half so this should be a matter of careful design and a minimum of economizing.

Ladder Rollers
A number of improvements have been made in ladder rollers including lubrication and bearings, seals and resistance to wear. Maintenance procedures are important to be established to prevent failure and extend the life of each unit.

Main Drive
The main drive consists of bull gears, pinions, intermediate gears with shafts, V-belts, and electric motors. The design of this system should correspond to the size of driven components and digging depth, nature of the ore and other factors that will result in digging forces. Electric motors, whether AC or DC, need to have variable speed provisions and usually through some form of thyristor/SCR electronics or Ward Leonard System for AC motors.

The design of drive evolved from a single motor and bull gear, to twin motors with two bull gears and synchronized drive of the upper tumbler. Problems with synchronization has resulted in higher amounts of power and motors, attempting to rectify the problems. At one time, successful dredges operated with a single motor driving the bull gear and buckets. That same motor was then used to alternately raise and lower the ladder, stopping the bucketline in the process. The economy and efficiency of that system, while requiring a "heads-up" operation according to past dredge masters working with them, should not be lost on future designers.

Winch System

A seven drum winch that combines bow, stern, spuds, and headline when used, has been an accepted standard in all types of BL/M dredges. With the advent of individual modular winches developed for shipboard and offshore drilling platforms, there are several optional systems though costly. A separate ladder hoist winch is normally used due to the heavy weight and control required.

Main Hopper

The main hopper receives the material dumped from the buckets as they pass over the upper tumbler. The material enters the revolving trommel from the hopper for washing and tumbling. A by-pass boulder chute, is usually provided above the hopper to prevent large boulders or other heavy material from damaging the trommel.

An actuated chute gate is used and controlled by the dredgemaster as he views the obstruction coming up the ladder. This material either bypasses the trommel and is carried aft on the overburden stacker, or, dropped by chute overboard at that point.

Trommel, Revolving Screen

The design of trommel screens has been the source of many mistakes and confusion. Standards and sizing dimensions are available from the sand and gravel industry to select screen area in relation to production volume. But in the many variations of mining conditions and soils, combined with clay and other problems that have to be overcome to wash out the target mineral and through the screen, other considerations must be taken into account.

Classification of feed to prepare for the concentration phase by mineral jigs, is an important function to be designed. Depending upon the nature of the ore or dredged material, whether containing clay, cobbles, boulders, organic trash, silt and sand, the primary consideration is to insure maximum throughput. The target mineral must be separated from waste in this phase and anything that deters this objective must be countered.

Some examples of trommel sizing in relation to bucket size is set forth in Table 5 Trommel vs. Dredge Size. This does not intend to imply however that these were the correct ones but at least the dredges mined successfully and therefore is some indication. Note: The last example includes two trommels, a practice that became popular in larger dredges in Malaysia; minimizing the diameter of a single trommel and maximizing screening capacity.

Table 5 does not show the slope used in each example but generally, it is 1:10 or 1:12. Again, there is some controversy on the slope to use; not too steep so as to give the material time to be washed and broken down, or not too flat so that the material does not back up in the trommel.

A position that has been taken by Cleaveland is that other means of keeping the material moving, such as screw conveyor baffles in the trommel combined with a flat trommel, would optimize throughput by increasing exposure time of the material to the screen holes.

This would tend to increase the through-put efficiency and thus, reduction in size. Also, he suggested that the sides of the trommel should be flat plates and various configurations. Octagonal, pentagon, 12-sided, have been studied. In these cases, the trommel would present a more stable surface for screening vs. curved and surging, thus increasing exposure time of undersize material to screen holes. Increased wear is one of the negative factors.

Screen Design

The normal procedure developed with BL/M dredges has been to use either high tensile steel or case hardened mild steel, plates, bolted into the periphery of the trommel frame and gang-drilled using tapered drills. Some plates have been cast with the holes in them. Later developments have used hard rubber plates with steel backing, molding the perforations and taper. Depending upon the abrasiveness of the material, this has produced significant benefits in some applications such as Malaysia for tin.

The selection of screen hole size is an important factor that recalls the basic objective of the trommel screen; <u>optimum throughput</u>. In earlier days of gold dredging before mineral jigs were accepted, a multiple flow line system from the trommel discharge or undersize passed the slurry into banks of riffles on each side. As many as 50-60 separate flow lines were used to direct the flow. It became apparent however that most of the material was flowing through the forward end of the trommel screen and thus overloading those riffles and starving the lower-end.

The solution to that dilemma took several turns. The basic solution was to reduce the size of the holes in the forward part of the trommel, to 1/4- or 3/8-inch and increasing the hole sizes in the lower-half; thus achieving more uniform flows into the riffles. The effect of course was to reduce the overall efficiency and throughput of the trommel. Another solution was the "Barker Splitter," using crossing flow lines and directing the upper flow into the lower riffles and visa-versa.

When mineral jigs were introduced using Bandeleri, Pan American, Yuba, Harz, to name the principal ones (see Chapter IV-Mineral Jigs), all were rectangular jigs with limited flow capacity in each. Thus the same or similar amounts of flow lines were required as with riffles and the solution for even distribution remained the same. However, once the Cleaveland Circular Jig was introduced the number of flow lines was reduced significantly. This caused some dredge designers to rethink screen hole size and to attempt to optimize throughput with smaller screen areas.

Mineral jigs of the Cleaveland Circular design, have been shown to function best with (-)1/2-inch material. A trend to used slotted holes to increase throughput without hole size, such as 3-inch by 1/2-inch has produced good results. Using rubber screen plates this shape of hole can be easily cast but steel or cast-hardened iron plates are effective as well. Some useage of hard rubber has shown the life to be three-five times that of steel plates.

Save-All System

In the process of the buckets stepping over the upper tumbler and discharging into the hopper and trommel, there is a certain amount of spillage; material that misses the trommel. Below the discharge point at about deck level, a set of grizzly bars are installed on a vertical slope to catch this spillage. The material screened by the grizzly bars at a selected spacing; i.e., eight to ten inches finds its way into the "save-all" recovery area.

Studies in California and Colombia with gold, and Malaysia with tin have shown that Save-alls have recaptured as much as 12% of production/recovery of the target mineral. A 6-7% average seems however to be the norm and depends upon the overall dredge design and operation. For that reason, most dredges have installed save-alls but not all have paid much attention to their cleanup and security.

The standard approach has been to use riffles and locking up the room in an area below the hopper discharge. Neglect in cleaning up results in the riffles becoming clogged and ineffective. It is also an area for "high grading," since it can result in nugget accumulation.

The most effective system for save-alls has included installing a small revolving trommel screen fed from the grizzly, with the undersize material dropping into a sump. That material is pumped into the rougher jig distributor. In that way, the system is "closed" without access for high-grading and is self-cleaning.

Circular Distributor

The trommel screen is the usual method of classifying undersize material with the intention being to reduce the volume that must be treated further by jigs. A design by Cleaveland of a Circular Distributor, is particularly useful where there is excessive amounts of clay and silt.

The Circular Distributor is installed under a single discharge point from the trommel casing (instead of multiple flow-lines), providing a splitting and overflow of material before entering the jigs. This system is new and has yet to be proven, but calculations by CPD, suggest that there would be substantial benefits from its use in some deposits.

Section 3
Mercury Amalgamation in Gold Placer Mining

Background
The careless use of mercury (Hg) in gold placer mining, is causing a great deal of concern. The problem has been dramatized with reports from Brazil claiming large quantities of Hg polluting rivers and streams from the mining activities of small miners (guarimpeiros). The recent Berlin Conference on Mining and the Environment (June 7, 1991 edition of Mining Journal), included a section devoted to mercury pollution in developing countries.

We have seen this occuring in other developing countries where gold placers are mined. The crudest methods of mining and recovery are followed, lacking the technical know-how or funds to procure proper mining, recovery systems and procedures. Considering the political implications, the possibility that governments will begin to enforce a prohibition against the use of Hg in artisan mining of gold placers much less the mining itself, is generally remote.

Our concern and hence this article, is that in the panic to right "some evident wrong," there might be a blanket restriction placed on the use of Hg in gold placer operations. Just as in the case of asbestos, the scientific evidence has shown the popular outcry to be largely false. I would therefore like to present the case for "safe useage" of Hg in gold amalgamation, while not ignoring the obvious dangers when carelessly handled.

Chemical Composition
Mercury used in amalgamation will be of commercial grade, which has a nominal purity of 99.7%. There are traces of copper, silver, manganese, iron, magnesium, lead, chromium, nickel and aluminum as impurities. The melting point of Hg is (-)37.97 degrees F; and boiling point of 675 degrees F.[1]

Characteristics

Hg is the only metal that is liquid at normal ambient temperatures, is silvery white with a faint bluish tinge. Below its melting point, it is a white solid and above its boiling point, it is a colorless vapor. Other properties are high density, uniform volume expansion, electrical conductivity, ability to alloy readily forming amalgams, high surface tension, chemical stability and toxicity of most of its compounds. A flask contains 76 lbs of Hg, the usual unit of measure. Most purchases of Hg by the government specify 99.9% pure.[2] It is insoluble in water.[3]

Method of use

Hg is used in the recovery phase of gold placer mining. When amalgamated with gold, it is placed into a retort to boil off the Hg. The system is closed to the atmosphere and the Hg vapors are condensed into a water beeker. All Hg is kept in steel flasks that are in turn, stored in a safe in the locked "gold room" on board the dredge or wash plant. A continuous flow of supplemental water should be maintained during operations over the exposed Hg on plates and auger riffles, to prevent direct exposure of Hg to the atmosphere. In particular, this will prevent ambient level vaporization during hot days.

Safe Systems

Our company has developed and applied procedures and systems for safe useage of Hg in gold placer dredging recovery over a period of years, since our founding in 1930. The use of these systems on our bucket ladder mining dredges (BL/M) as well as portable wash plants, was developed principally in Colombia (Pato Consolidated Gold Dredging, Ltd.) and New Guinea (Bulolo Gold Dredging, Ltd.), from our company in California. There was also a close interaction on technical developments in those earlier days, between our companies with Yuba Consolidated Gold Fields and Natomas. Too many of the dredge owners however, held onto riffles in their recovery systems even though jigs had proven their superiority.

Before BL/M dredges were equipped with mineral jigs for recovery of gold, it was a normal practise to spread Hg over riffles to help catch finer gold. Attempts were made to reclaim the Hg with "traps," but it is known that some found its way into the gravels and rivers. This is evident today when BL/M dredges mine through tailings of earlier dredge mining and become "net producers" of Hg, along with gold.

As mineral jigs were installed on dredges, replacing riffles and coupled with amalgamation systems that controlled Hg escaping into the environment, there was no longer a need for "open area" use of Hg. The "Jackpot Amalgam System" was developed and used on our dredges. An Amalgam Barrel was used for many years on Yuba dredges. Our company tested it however and found it less efficient than the Jackpot system and thus did not adopt it on our dredges.

Recovery System w/Hg
The system that uses Hg should include mineral jigs in order to produce a concentrate that is of less volume and directly amendable to amalgamation, in a closed circuit. This assures recapture of the Hg before discharge of water and concentrates to tailings. Such a system can be scaled down in size from large volume dredging operations but cannot tolerate open riffles or flumes in a continuous process.

For instance, the use of riffles as in a "Long Tom" can be done on a batch basis, applying careful procedures and preventing vaporization of the Hg. This can be a fall-back position for the small miner but should not be confused with a commercial-scale, mining project. The following recovery system therefore is illustrative of a continuous, gold placer dredge mining operation.

Mineral Jig Function
The primary purpose of a mineral jig in the recovery process, is for separation of waste materials for discharge to tailings, and concentration of the heavier minerals for further processing. Jigs can handle a wide range of material compositions with minimum care and main-

tenance; a basic characteristic of most placer deposits. This reduces the volume of concentrates for processing in the final phase of recovery with Hg. A primary and secondary level of jigging when properly designed and maintained, has proven to be sufficient for gold placer recovery in most cases.

That is, there is no need for further jigging even though there has been a tendancy by some operators to add more jigs as "insurance." At the same time, the operation of mineral jigs is not well understood within the mining industry today. This may account for the neglect of jigs or substitutions by other, less efficient equipment.

The primary jig phase should result in about 90% of the feed being discharged through the launders as waste to tailings. The other 10% becomes a concentrate, as the hutch product which is fed to the secondary jig for further processing (i.e. commonly referred to as "rougher and cleaner" jig phases).

A feed of (-)1/2 inch material has been found to produce optimum performance by the primary jig. Feeding of finer material to the jig will tend to increase the ratio of concentrate as the hutch product. The proper performance of jigs requires "tuning" of the system including stroke and cycle of the drive, ragging depth and active raking of the bed on a periodic basis, to mention the basics.

The secondary jig phase treats the primary jig hutch product. The resulting concentrate is then fed directly to the Jackpot Amaglam System. At no time in the jigging process should any Hg be used. If as mentioned, dredging is being done in early gold fields and tailings, Hg may be found in the hutch products of both primary and secondary jigs.

Jackpot Amalgam System

The system is designed to expose the Au to Hg in three stages, shown in Figure 2. The first is the Jackpot which is a cone designed to dewater the secondary jig discharge or hutch product, producing a boiling effect with the Hg. Below the cone is a pot containing Hg. From the lower slopes of the cone are flow lines, directed to the leading edge of the plates. A majority of the Au is trapped in the Hg pot and periodically removed through a discharge hole that is plugged at the bottom.

The overflow of concentrates and water through the tubing from the cone, flows first over "auger riffles" filled with Hg. The "auger riffles" are neoprene rubber, cast with round holes about 1/2-inch in diameter and vulcanized to a steel plate. Following the auger riffles the concentrates containing Au that has not yet amalgamated, flow over silvered-copper plates coated with Hg.

At the lower edge of the plates, is a discharge launder sloping into a Hg trap. The overflow of concentrates is then either discharged into the tailings or reprocessed through a Ball Mill. In the latter case, this is used to polish any tarnished Au that may be in the deposit, followed by another amalgam plate for final recovery before the Hg trap and discharge of waste to tailings.

Handling of Hg

While there are detailed handling procedures for cleaning up the amalgam system, one of the essential guidelines for protection of personnel is to keep a constant flow of water over the Hg when it is exposed to air. Always wear rubber gloves when handling Hg in the cleanup of the jackpot and plates.

A typical check-list of procedures in handling cleanup of amalgam, is listed as follows but will vary with the size of the operation, production and laboratory facilities.

-Insert small plug in bottom of mill. Soap threads to get a good seal.

-Remove top and side covers and dump in amalgam.

-Start filling mill with water. Add 1/2 oz of Hg for each oz of amalgam.

-Put in rods. Put on side cover. Stop water when mill is 3/4 full. Put on top cover.

-Grind for 5 to 15 minutes. Grinding time varies with quantity of amalgam and size of gold. Excessive grinding time causes adsorption of Hg.

-Remove top and side cover and rods.

-Drain water and amalgam through small plug into pail. Bail and sponge off water. Skim off as much Hg as possible into another pail.

-Put 1/2 cup of amalgam into Hg. Agitate Hg and amalgam with water hose and moderate stream of water, pressing hose against bottom and sides of pail to float off dirt. A circular motion starting from center and spiraling to outside edge produces best results.

-Bail and sponge off water and dirt. Repeat agitation, bailing and sponging until clean; two or three times is usually sufficient.

-Pour clean amalgam into a canvas cloth placed over empty pail and squeeze dry. Place resulting ball of amalgam in container to drain; pound ball into compact mass in bottom of container.

The procedure above is repeated to insure recovery of Hg and cleanup of system.

When the amalgam is extracted from the system, it is placed in a sealed retort for boiling off of the Hg. The Hg vapor passes through a condensor pipe that cools and discharges liquid Hg into a sealed, rubber container that is kept moist with water. The Hg is later reused in the amalgamation system and is kept in sealed containers in a safe. This process also requires detailed steps for both safety and security purposes.

Summary
While this has been a general description of Hg use in amalgamating and recovering gold from placer deposits, the key message is that it is and can be accomplished, with minimum danger to personnel. It is the most efficient method of recovering gold from placers in large and small mining operations. However, guidelines must be established and followed based upon proven experience. CPD provides those guidelines and training on a contract basis.

References
[1] "Metals Handbook," Vol. 1, Properties and Selection of Metals, Publisher American Society for Metals, pg. 1215.

[2] "Mineral Facts and Problems," U.S. Bureau of Mines, pg. 669.

[3] "Chemical Engineers' Handbook," McGraw Hill, pg. 312.

Section 4
Other Bucket Dredges

Bucket dredges other than BL/M, have their applications that are appropriate to mining. The bucket clam (BC), grab (BG), dipper (BD) or dragline, are all common to some form of mining (see Figure P11.0). Prior to the advent of the backhoe for small scale placer mining, the dragline was the popular tool; feeding a floating wash plant. This system was called a "Doodle Bug," and during the 1930's, there were reported to be as many as 350 sold in California alone (a company named Bodinson, San Francisco, was the largest producer).

Clamshell/Grab Bucket-(BG/C/M)
Applications of clamshell or grab dredges mainly for S&G, have been suggested for placer deposits. Such measures have been attempted in the past but for similar reasons to the suction dredges, compounded with its limited movement as well as inability to clean bedrock thoroughly, this system cannot be recommended for precious minerals.

One of the difficulties introduced with bucket systems that cannot feed the processing plant direct, is the consequent rehandling of material that is required. The result is higher costs of operation and a lack a security of the mineral being mined. Reclamation when required is also made more difficult and costly.

Dragline
An example is in using a dragline which tends to be specified in large volume, such as 4-10 yd^3. Such a large amount of material cannot be dumped directly into a processing plant since it will choke the classification process. The dragline is not effective at digging and thus cannot clean bedrock thoroughly.

The result is stockpiling and working from that pile; exposing high grade material to theft and requiring additional handling. The backhoe development was the salvation for small to medium sized mining, using a washing plant and replacing the dragline.

Backhoe-(BB/M)

The best development for small scale mining of placers including gold, diamond and tin has been the backhoe. While normally limited to volumes under 100,000 yd³/month when keeping the bucket size to a manageable dump load (i.e., 1/2-2.0 yd³), it can be sized to maintain a fairly steady flow of material to the trommel and processing plant.

Digging from the bench adjacent to a floating washing plant, it can dig above or under the water, scraping and cleaning bedrock fairly efficiently. While cost of operation ($/yd³), will normally be two to four times that of a BL/M dredge, its capital cost and speed of mobilization is attractive to the small miner.

The processing plant for a BB/M system can be simply a scaled-down version of the BL/M system. The plant can be mounted on skids and towed from point to point when mining in the dry though handling of tailings and processing water becomes more of a problem. It is preferable to float the processing plant so that it can discharge tailings behind, filling and reclaiming the land in the process.

Unfortunately, many systems of this kind are still used with a simple, flume-riffle recovery system. The discharge is usually accompanied by a great deal of sand and clay that will cause environmental problems; mud in the streams for instance. The use of a mineral jig system on the other hand, can permit recirculation ponds and means of controlling discharge that can accomodate environmental objections. A more thorough discussion of this system is in Chapter II.

Continuous Dragline Dredge

A concept of a new type of placer dredge mining system was invented in 1967, by Norman Cleaveland, which he called the "Continuous Dragline." The idea came from his observing a dryland, canal digging machine in operation in California, developed by Guntert & Zimmerman. Cleaveland invited Roland Guntert and another famous placer mining engineer, Patrick O'Neill, to join him in development of the system, which they named the "COG," using the initials of the three partners.

Guntert built a scale model prototype in a warehouse at his plant in Stockton, California. Tests were successfully run. Attempts were made to sell the idea to mining companies over the next several years, but did not succeed and to this day, remains as a potential for large volume and deep water placer mining (see Cleaveland 1967).

Section 5
Bucketwheel Suction Dredge (BWS/M)

Development of the underwater, bucketwheel suction system for dredging began in the early 1960's. The dry land bucketwheel systems used in massive open pit mines around the world for years prior, provided the basic technology, on which to build the submerged systems (see Figures P19.0, P20.1).

Perhaps the best application to date for these machines, has been in mining for cassiterite, rutile, zirconium sands, ilmenite, magnetite, and sand & gravel. I am not aware of any successful applications for placer gold, one of the problems being its difficulty in uniformly cleaning bedrock, or, a smooth surface.

Some of the BWS/M dredge's best features include lower capital cost (than a BL/M), portability, higher concentration of solids versus other suction dredges (i.e., 30-40%), and higher relative production to capital cost. The speed with which the dredge can be mobilized and transported, are other advantages that appeal to some miners. It is basicly "new" with recent developments improving on its efficiency and solving some of the operating problems. This generally attracts the current craze for high-tech in contrast with older systems such as the BL/M dredge.

At an Alluvial Mining Conference in London (November 1991), considerable discussion ensued questioning whether BL/M dredges had been given fair consideration in selecting new alluvial mining systems particularly with industrial minerals. The trend has been towards BWS/M and spiral concentrators, for instance, in industrial mineral applications without giving alternative analysis of BL/M systems and their cost-effective benefits.

Some of the problems inherent in the BWS/M system need to be considered in contrast with other systems. These problems include:

1. Considerable water introduced into the system that must be dealt with in dewatering at the processing plant (70% water typically).

2. Digging on a radian of arc requiring small steps forward and redredging areas to avoid passing over furrows of pay material; high wear with hard or abbrasive material, cobbles or boulders which it cannot pass.

3. The suction system behind the buckets, producing restrictions for clay, jungle trash and larger materials (that are routinely scrubbed in BL/M systems up to 12" or more in diameter), with downtime to clean out the system for obstructions and worn bucket teeth.

In summary, there are excellent applications for BWS/M systems but there needs to be a balanced, engineering and cost analysis of the tradeoffs with BL/M and other systems. The ruggedness of the operation, continuous operation, environmental considerations of reclamation of tailings and discharge of slurry and excess water, R/E (recovery versus evaluated content of the deposit), ultimate cost of extraction including amoritization, are some of the important considerations.

Section 6
Other Suction Dredges (CS/M, S/M, TH/M)

Cutter Suction (CS/M)

The largest population of suction dredges is Cutter Suction (CS) or, (mining application) CS/M. This type of suction/hydraulic dredge is characterized by some form of cutter, usually a "basket" cutter head, with a suction intake behind and at its base. The cutter motor provides the digging force and a pump is located in the dredge barge and often is combined with a submerged pump mounted on the ladder (see Figures P16.0,P20.2).

Ladder Pump

The benefits of the ladder pump were developed using hydraulic power, in the early 1960's, resulting in subsequent retrofits of existing dredges and installations on new ones. Production and concentration levels are dramatically increased with their use.

Instead of a suction pumping operation from the hull, with concentrations or solid ratios of 15-20%, the submerged ladder pump has increased that to about 30% with the same or less horsepower; essentially a positive displacement pump. The installed pump in the dredge then becomes a booster pump for transport to the discharge or disposal area. In the case of mining, to the processing plant.

The problems of the BWS/M dredge in mining are amplified to a greater degree with the CS/M dredge. Without the benefit of positive displacement of the digging wheel buckets, the suction process is even less effective. For precious minerals the suction mouth of the pipe produces a turbulence that disperses fine particles such as gold or diamond, and cannot clean bedrock effectively.

Plain Suction Dredge (S/M)

Mining with S/M dredges is in fact the most plentiful though usually small in size, throughout the world (see Figures P15.2, P22.0). The S/M system has been effective in such areas as Luzon, Philippines for magnetite beach sands using multiple, small units. But when used for precious minerals such as gold and diamond; in South America, SE Asia, Africa, numbered in the thousands of units they are an example of mineral waste. It is the common tool of the "illegal miner," and is usually handled by a diver directing the hose into the sands and potholes. Recovery is normally by a flume/riffle system mounted on the barge, with perhaps 25% recovery of the in situ minerals.

Due consideration and recognition should be given to the S/M dredge for its beneficial applications. For years using a submerged pump, it has been effective for sand-winning in The Netherlands and elsewhere at great depths (over 200 feet). It is used in many areas of the US for sand and gravel dredging where the material is free-flowing. The airlift pump though not using a suction principle, has been used in similar applications with some success though highly energy consumptive-per-volume dredged.

Trailing Suction Hopper dredge (TH/M)

This type of dredge in a mining function, has principally distinguished itself in sand and gravel winning in the English Channel (see Figure P17). The TH dredge is a self-propelled ship-shaped vessel and was developed in the USA for river navigation dredging. It has a number of mining applications that may be exploited in the future, either in large rivers or in the ocean.

Some viable applications in addition to S & G for the TH/M dredge, would be in heavy mineral sands including deposits of rutile, zircon, ilmenite, magnetite, sand & gravel, and phosphates. It is not likely as a prospect for precious minerals where a bedrock must be cleaned for some of the same reasons cited above for other suction dredges.

Section 7
Electronics for Dredge Mining

Automation-"Fuzzy Logic"

With the introduction of micro-circuitry and compact computer systems, automation has been introduced into several types of dredges. For conventional dredging, perhaps some of the most advanced work was reported in WODCON XII in Orlando, Florida in 1989.

The paper reports research work done in Japan, adapting "Fuzzy Logic" systems of mathematics originally developed by Professor L.A. Zadeh, 1965, at UC Berkeley, California, USA (see Miyake/ Ofuji et al). The method has been integrated in computer technology programing called "Artificial Intelligence" (AI) into what are called "Expert Systems" programs.

Quoting from the cited paper, "A fuzzy control system, however, integrates the expertise of an experienced operator. That is, fuzzy logic, which enables a computer to make quantitative judgements in the same manner as a human, is built into the computer control system to offset the demerits of a conventional system. The system is thus able to treat fuzzy judgements like 'a little bigger' or a 'little to left' based on the know-how of skillful operators."

Conclusions of the study and extensive experimentation involving 70 professional engineers for five years from 1983-88, are summarized as follows:

"Initially, 244 control rules were prepared for the controls for dredging production, cutting pattern and forward shift. For the purpose of the field tests, however, these were reduced to 69, due to the sea conditions and soil conditions. It has hitherto been considered difficult to automate dredging operations. Our system, however, demonstrates the feasibility of fully automated dredging of efficiency comparable to that achieved by a skilled operator. In the future, fuzzy control will be applied not only to cutter suction dredges, but also to other work vessels. In addition, this automatic operation capability can be used to record and store important operational expertise of experienced operators, which can then be used in the training of inexperienced operators."

Placer Mining Applications

There are not many differences between the dredging problems of a cutter suction (CS) dredge for conventional dredging, BL/M or other type of dredge for mining. Automation has been adapted perhaps more in the large tin dredges in Malaysia than in gold but mainly in an automatic pilot concept, with an operator standing by to take over.

The fact is however that when the next BL/M dredges are designed, "Fuzzy Logic" systems in Japan should be analysed carefully for applications. It appears that much of the research has been done in Japan by the major dredging companies, builders and government in concert and they should be approached for technology transfer.

With the limited availability of experienced personnel in BL/M dredge mining, the concept of training personnel, passing on knowledge or techniques is especially appealing. One of the problems confronting placer mining today is the lack of experience and the resultant mistakes made in equipment selection and operation.

Control Systems

The electronic control systems and displays available today, stemming from development in conventional dredging with CS and TH dredges in particular, are adaptable to dredge mining. Various dredge builders and electronics suppliers have modern systems that can display positions of the ladder, swing, and area dredged. The push button controls for all functions replace mechanical systems. However, it is not the intention of this book to treat this aspect since it is generally one of control system hardware that is commercially available.

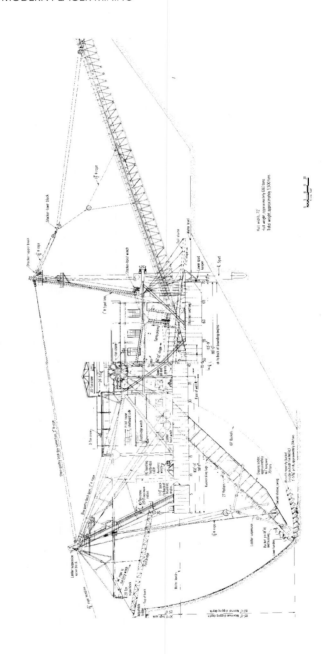

FIGURE 1-ELEVATION OF 18 FT³ YUBA DREDGE NO. 110, 85 FT DIGGING DEPTH

FIGURE 2 - JACKPOT AMALGAM SYSTEM
SCHEMATIC DIAGRAM

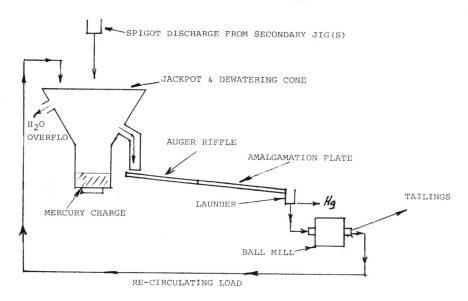

FIGURE 2-JACKPOT AMALGAM SYSTEM, SCHEMATIC DIAGRAM

TABLE 1
BL/M DREDGE PRODUCTION

CAL-TYPE BL/M DREDGES	BUCKET SIZE-FT^3	MONTHLY-YD^3
PATO #6 (CPD PARENT), COLOMBIA	2	67,167
ROARING RIVER, CALIFORNIA	3	60,000
YUKON CONS.GOLD CORP, DAWSON#7	5	75,450
SOAMER.GOLD&PLACERS(SAGP)#1, COLOMBIA	5	142,380
YUBA, MONTANA	6	149,114
YUBA MERCED #2, CALIFORNIA	6	165,000
BIGGS #4, CALIFORNIA	6	94,500
PATO #1, COLOMBIA	6	201,750*
USSR&M #5, ALASKA	6	179,700
" #6, "	6	168,000
" #8, "	6	151,200
YREKA GOLD DREDGING CO., CALIF.	6	145,000
SAGP #2, COLOMBIA	6	171,990*
YUKON CONS.GOLD CORP. #5, CANADA	7	149,000
" " " " #6 "	7	108,000
SAGP #2, COLOMBIA	7	217,980*
" " "	7.5	267,920*
GOODNEWS BAY, ALASKA	8	175,650
SAGP #6, COLOMBIA	8	269,010*
" #4 "	8	263,340*
PATO #2, COLOMBIA	8.5	167,750
CAPITAL #1, CALIFORNIA	8.5	215,400
" #2 "	8.5	223,170
SAGP #4, COLOMBIA	9	156,870
YUBA BIGGS #1	9	303,776
" " #2	9	319,778
" " #3	9	328,469
CALLAHAN, CALIFORNIA	9	250,570
YUBA MERCED #1, "	9	234,974
JUNCTION CITY "	9	240,000
LA GRANGE, "	9	311,431
SAN JOAQUIN, "	9	432,945*

TABLE 1 BL/M DREDGE PRODUCTION

USSR&M #2, ALASKA	10	174,000
" #3 "	10	263,100
NATOMA #5, CALIFORNIA	11	356,200
" #6 "	11	387,825
" #8 "	11	426,610
SAGP #6, COLOMBIA	11	406,350
SOUTH PLATTE, COLORADO	11	388,276
PATO #7, COLOMBIA	13.75	450,167
" #3, "	13.75	458,250
" #5, "	13.75	472,750
" #4, "	13.75	476,750
" #9, "	13.75	550,000
YUKON CONS. GOLD CORP.#2, CANADA	16	292,000
" " " " #3, "	16	400,000
" #4	16	271,000
YUBA #14, CALIFORNIA	18	368,746
" #15, "	18	383,334
" #17, "	18	382,477
" #18, "	18	425,330
" #19, "	18	418,760
" #20, "	18	379,292
CAPITAL #3, "	18	427,576
" #4, "	18	457,505

MALAYSIAN TIN DREDGES/EURO-TYPE

BURJUNTAI TIN #9	22	610,000
PER RIOTINTO #1	24	800,000
KUALA LANGAT MINING #1	30	850,000

*Hi-Speed bucketline

TABLE 1 BL/M DREDGE PRODUCTION

TABLE 2

SEG.NO.	DESCRIPTION
00A	Engineering
01	Hull, steel
01S	Truss & Gantries
02	House Framing
02D	Stairs, Platforms
03	House Covering
04	Buckets and Bushings
05	Bucket Pins
06	Bucket Idler
07	Lower Tumbler & Shaft
08	Upper Tumbler & Shaft
09	Ladder, Digging & Castings
10	Ladder Rollers, Bearings, Shafts
10A	Lubrication System
12	Ladder Suspension
13	Ladder Hoist Winch
15	Main Drive Gearing System
18	Winches: Bow, Stern, Spuds, Headline
19	Sheaves & Fairleads
20	Operating Controls & Console
21	Main Hopper
22	Revolving Screen(s) & Drive
22K	Save-all Screen & Drive
24	Distributor
24E	Casing, Revolving Screen
25	Stacker Hopper & Rock Chute
26	Stacker Hoist Winch
27	Stacker Ladder & Seating
29	Stacker Drive
30	Stacker Suspension & Side Guys
31A	Gold Room Housing/Screening-Security Room
31D	Amalgamation System

TABLE 2 BL/M DREDGE SEGREGATIONS

33	Tail Sluices & Supports
34	Grizzly & Save-all
35	Mineral Jigs Systems
35D-L	Seatings, Launders, Concentrate Sump
35R	Jig Steady-Head/Gravity Water Tanks
36	Spuds (if used)
37	Spud Suspension
39	Back Guys & Connections
40	Wire Rope
41	Belting, V Belts, etc.
42	Cranes, Trolleys, Hoists
43	Ladder Service Platform & Footbridge
43B	Service Platform Hoist
43C	Auxiliary Hoist
44	Tools, Grease Guns & Workshop
45	Gauges (instrumentation)
46	Valves, Check
47	Pumps, Water Service
47E	Pumps, Slurry
48	Piping, Water
48E	Piping, Slurry
48G	Bilge & Ballast System
51	Paint & Lubricants
53	Safety Guards
54	Communication Systems
55	Electrical System; Motors, SCR, Wiring
55A	Transformers
55E	Power Cable & Entrance
58	Diesel Generator, Auxiliary
58A	Diesel Generator-Main Power(if no shore power)
80	Compressed Air System
82	Spare Parts Provisioning

TABLE 2 BL/M DREDGE SEGREGATIONS

TABLE 3

BUCKET SIZE FT^3	PRODUCTION VOLUME	
	YD^3/HOUR	YD^3/MONTH
6	260	187,200
9	390	280,800
12	520	374,400
14	607	436,800
18	780	561,600
20	867	624,000
24	1,040	748,800

TABLE 4 BL/M DREDGE HULL SIZE COMPARISONS

BUCKET SIZE FT^3	HULL-FT			DISGING DEPTH FT
	LENGTH	BEAM	DEPTH	
6	100	48	8	30
8.5	125	46	9	48
10.5	168	72	10	85
14	171	66	11	91
16	221	72	12	95
18	251	80	11	124
20	200	93	12	85
24	325	110	14	150*
27	300	94	13	130*
30	376	97	14	148*

*European-type; Malaysia tin dredges

TABLE 5 TROMMEL VS DREDGE SIZE

BUCKET SIZE FT^3	TROMMEL SIZE-FT		
	DIAMETER(ID)	LENGTH	SCREENING LENGTH
6	6	32	22.4
8-1/2	6	37	27.0
10.5	7	38.6	26.6
13.5	8	48	33.5
20	9	50.5	37.1
24	9	68.5	51.1
27	(2) 9	59.6	43.3

TABLE 3 BL/M DREDGE SIZE VS PRODUCTION TABLE 4 BL/M DREDGE HULL SIZE COMPARISONS TABLE 5 TROMMEL VS DREDGE SIZE; BL/M

CHAPTER IV
MINERAL JIGS

Introduction

Mineral Jigs have not received wide useage in the placer mining industry in recent years considering numbers of mining units and activities on a small scale. For that reason jigs are not well understood by many in mining. As a result, some of the newer operations in placer gold mining have reverted to inefficient riffle systems, losing large quantities of gold in the process. Likewise, there are hard rock mines of various minerals that either have old jigs installed but the personnel are not familiar with their use and maintenance or new mines where jigs could be of advantage but no one is suggesting them.

With the above needs in mind, I compiled and Consolidated Placer Dredging Inc. published, "Handbook of Mineral Jigs," in 1983. This chapter is an update of that material and summarizes the history of mineral jigs from inception to the current day.

In delving into the records of mineral jig development, I found that the engineering and experimentation conducted during the late 1800's to the early 1900's, was thorough and professional. Thus it can provide useful data for applications of mineral jigs to contemporary mineral recovery problems.

The slow acceptance of jigs on dredges, the skeptical resistance that held to riffles or their combination with jigs, is surprising. But tracing that development it will be shown that the leaders in gold and tin dredge mining, eventually made the switch and benefitted by doing so.

Section 1
Milestones of Mineral Jig Development

1556-Report of "circular sieves" being used as a Mineral Jig for placer gold recovery in Europe (*De Re Metallica*).

1800's-Development of hand and power-operated jigs for mill operations in lode mines; i.e., coal, ironcopper, gold, tin, silver, zinc, etc.(*Ore Dressing,* Richards).

1912-First jigs(Woodbury)installed on a BL/M dredge mining GOLD; Merced River, California; design and construction of the first jig for dredges, by J.W . Neill. Tested on that dredge(*Yosemite No.1*).

1922-First jigs installed on a BL/M dredge mining TIN, in Malaya, using Harz-type jig, by Pacific Tin Consolidated Corporation.

1926-Design and testing of "Crangle" jigs on BL/M dredges; Pato Consolidated Gold Dredging Co.,Colombia. Testing program begun by Frank Griffin with team of Natomas engineers experimenting with jigs on BL/M gold dredges in California.

1936-Replacement of riffles by jigs on dredges (Idaho, Calif., Nechi River, Colombia; Bulolo, New Guinea.) Design/development of PanAmerican Jig by Placer Management Ltd.(CPD predecessor).

1947-Design/development of Yuba jig, Yuba Consolidated Gold Fields Corporation.

1953/67-Design development and patenting of Cleaveland Circular Jig; first successful use of circular jigs on BL/M dredges; for tin in Malaysia and diamond in Brazil.

1968/70-Testing of Cleaveland Circular Jigs on a BL/M Gold Dredge (Colombia - Pato Cons. Gold Dredging Ltd.), alongside Pan-American jigs. Led to first installation replacing PanAm jigs, in 1974.

1979/82-Development and patenting of Mk II-Cleaveland Circular Jig; improvement on 1967 design.

Section 2
History of Mineral Jigs

While this discussion will tend to focus upon the applications of mineral jigs to placer gold, tin and diamond, the subject would not be complete without a review of the origins of jigs. Their uses and theory since early testing and development, revealed a great deal that could serve to enlighten us today in their applications and use.

It is commonly accepted that the first published document on mining technology was, *DE RE METALLICA* by Georgius Agricola, published circa 1556. In this book which was a product of the author's travels among the mines of Europe, he stated that "a recent practise" particularly for the purpose of recovering alluvial gold, which was being sluiced from streams and rivers, was "the use of basket sieves."

Figure 1 Circular Sieves in 1500`s, is a drawing taken from De Re Metallica.

The circular basket sieves as indicated by the drawing, were worked up and down in buckets of water; forcing the water upward through the sieve and the contents, then lifted out of the water to allow gravity to pull the heavy minerals down through the lighter fractions and into the tub. The finest sieve was achieved by using hair as the screen to strain out the smallest particles of gold.

Lode Mine Jigs
The next phase of jigs is reported in "Ore Dressing" (Richards, 4 Vol's). In two of the volumes he reports on developments of mineral jigs in the 1800's, where they were used in mills for coal, iron, copper, zinc, lead, pyrite, barite and silver. Considerable engineering and experimentation took place during that period, developing the "laws of jigging." Much of this data is useful today since some of it may have been "lost" from the mining community in the passage of time.

There were two basic categories of jigs in wide use in the 1800's: The Hand-Operated and the Power-Operated. In large plants it was the accepted practise into the early 1900's, that one man was required to operate each jig, mounted in long rows in the mill. Each man would look to the maintenance, repair and cleaning of his jig to assure its proper functioning.

Power Jig
The leading power jig of the day was the Bradford Eccentric Jig. It appears that coal jigging led this development both in Europe and the US. The power jigs were sub-divided into two main categories: Moveable Sieve, and Fixed Sieve.

Moveable Sieve Jigs
The main types were: Conkling, Hancock Vanning, Bilharz Circular, Robinson, Hoopers Vanning (used for garnets) and the Cornish.

Fixed Sieve Jigs
 This type was dominated by the Harz-type Jig which was more widely used   for coarse and fine materials. Others that were used were: Collom(with modifications by the Dudley and Evans); Hodge, Parsons & Fisher (a Harz modification for iron ore); Bilharz-circular (used for slimes especially); Baum (coal); and Francon.

Hand Operated Jigs, Fixed Sieve Type
Utsch (coal); "Under-piston" jig; Diescher; Siphon Separator; Ferraris; Stutz; Sheppard; Luhrig; Osterspey; Henry Faust.

Section 3
Laws of Jigging

Factors and proportions for selecting and regulating jigs for particular minerals and ores include the following:

Stroke Length of Plunger
Increased length with the size of grain.
Flow rate of feed proportional to increase.
Density of ore was inversely proportional.
Depth of ragging on bed caused proportional increase.

Capacity of Jigs
Ratio of feed in volume-per-square foot of jigging area or sieve/screen
bed area.

Side-wall effect
Rectangular, or, trapezoidal jigs; causing friction and resultant dead
or ineffective areas of jigging, having a reducing effect on capacity.

Efficiency degrades with distance
From point of feed to bed, heavy minerals flowing to hutch in initial
area and less particles being affected with same power near edge or
discharge lip.

Overloading
Diminished efficiency of jigging process; under-feeding wastes energy and time. Compensating through increased stroke or speed for
overloading, causes boiling on areas of the bed that degrades concentration.

Size of Feed
Affects capacity; coarser grains can be fed in higher volumes than fine
grains. This effect is based upon ratio of grain size to depth of bedding
which affects the distance the mineral must travel down the bed to the

hutch. Early experience showed ranges of 3-1/2 inches down to 1/32 inches grain size. The former was in coal and varies with the mineral and ore.

Density of mineral
The higher the density the larger the volume of feed that is possible and conversely. The size or shape such as flat vs. round, as well as the coefficient of friction of the mineral can change the above relationship slowing down the rate in the former and speeding it up in the latter, regardless of density.

Hindered Settling
This is the phenomenon of mixed sizes, shapes and gravities which are sorted in a rising current of water, producing a higher velocity than the settling current. If a grain above has a greater density it will settle faster below the grain of less density. The largest grains will work to the top and smallest to the bottom. Thus, the upward pulse of the jig raises the material at a faster rate than gravity hindered by the friction and density of water will allow it to fall and results in the concentration effect of the jig. The horizontal velocity of the feed water over the bed, will carry the lighter fractions to the discharge lip.

Jig Screens
In early practice with lode ores, they used hole sizes varying from 0.69 to 31.75 mm, or, longitudinal spacing in the jig screens. The total area of the openings varied from current theory but this was no doubt caused by the different minerals involved, with a total open area ranging between 11 and 82%. Various materials were used for screens such as: Brass cloth; steel plate-punched holes; white cast iron plate; copper cloth; steel cloth; cast iron grating; iron cloth.

Principles Of Efficient Jigging
Even distribution on jig bed of ragging and feed material.
Uniform feed to bed; not erratic high to low.
Uniform ragging distribution.
Flow rate vs. distance to discharge lip.
Particle size vs. distance to lip.

Section 4
Placer Mining Applications

First Jig Installation on BL/M Dredge-1912
The first on-board Mineral Jig for a BL/M dredge was designed and installed by an engineer named, James W. Neill, circa 1912, Yosemite Dredging & Mining Co., operating on the Merced River, California.

The "Yosemite Dredge No. 1," was a 3-1/2 ft^3 BL/M dredge, and the engineers had been conducting a series of tests to determine what gold, if any, was being lost. These tests were made by securing samples from the fine material passing over the tables and also by driving pipes through the "fines" to bedrock at close intervals across the pond.

The tests confirmed that there were losses being experienced from the riffles and this led to the initiative to construct a small jig. Considerable experimentation followed with modifications being made to effect improvements in recovery.

Once this was completed to the satisfaction of the dredging managers, they ordered eight "large jigs of the Woodbury type" which had been developed and used on hard rock ores. This installation was only partially successful because of limitations of the jigs, lack of understanding, or, priority for maintenance and only being installed on a part of the sluices after the riffles.

Second Jig Installation
With the building of a second BL/M dredge, "Yosemite Dredge No. 2," Neill designed a new, improved jig (called a "Neill Jig") and placed ten of them at the lower ends of the first five tables of riffles. They were operated by eccentrics running on shafts at the side of the dredge. They used centrifugal pumps to transport the concentrates from the jigs to the settling boxes and amalgam plates.

Tin Dredge Recommendations
In 1916, Neill recommended consideration be given to installing jigs on dredges mining for tin, placing them in series. They were still evaluating results, but had verified improvements in recovery of gold from use of jigs and this with very clean gravels and sand. Neill's judgement was that where tarnished gold and heavy clays would be present that jigs would be even more important in saving gold.

Development of Dredge Applications

Natomas Tests
Following Jim Neill's jig development in 1912-14, a group called the Lyons Syndicate, financed a test program of jigs on a Natomas dredge. Those in the Syndicate were Frank Griffin (founder of CPD), his brother Maurice, and O.B. Perry, all members of the Natomas Co. These tests were conducted at Natoma, California, to evaluate the amalgamation of gold using Zinc amalgam, which failed.

The additional purpose of the tests was to evaluate the benefits of the Neill Jig in handling clay deposits. This resulted in the discovery that some of the placer gold was tarnished and thus would not amalgamate in that condition. They introduced a ball mill in the circuit to polish the tarnished gold and recirculated it through the amalgam system. This resulted in the design of a circuit using a jig with the riffles, followed by a ball mill before flowing to the amalgam plates.

The designers of the system were Frank Griffin, L. D. Hopfield and Edward Strouse, all of the Natomas Co. They commented upon the fact that it might eventually become possible to do away with riffles entirely, using only jigs for recovery, with the ball mill and amalgam system.

Pan-American Jig Development

The ensuing years were occupied with the above persons carrying on further experiments on jigs in placer dredging operations. This included the activities of an engineer working on the Pato dredges in Colombia, named Crangle. He developed the Crangle jig and tried it on the dredges in the 1920's and paved the way for the Pan-American Jig that was developed by Pato's affiliate after 1932, the Griffin Co. (CPD).

The Bulolo Gold Dredging Co., beginning in 1930, used the Griffin Co. to design and install eight new gold dredges in New Guinea. In the process of design along with Natomas Co. (where he was a Director), Griffin and his engineers began experimenting with the Bendelari jig in installing it on the first of the dredges to operate at Bulolo.

This installation proved the effectiveness of treating the entire production of the dredge with jigs. Out of those tests, Griffin's engineers designed the Pan-American jig and filed its patent, installing it in place of the Bendelari jigs on the Bulolo dredges in New Guinea, and Pato dredges in Colombia.

The Neill jig had also been tried in the early 's by dredge operators in Alaska and the Yukon Territory of Canada, but without a great deal of success. This may have been partially the result of their retaining riffles ahead of the jigs plus improper evaluation of the results. However, they were dropped for the immediate future.

The Bendelari jig was probably developed in the State of Missouri on lead-zinc ore mines, sometime in the 1800's. The U.S. Smelting, Refining & Mining Refining Co. (later renamed Alaska Gold Co.), adapted jigs more successfully later in the 30's, but only installed them on some of their dredges (possibly a conflict with "oldtimers").

Replacement of Riffles by Jigs

A notable installation of jigs, possibly the first that fully replaced riffles below the trommel screen distribution, was made on a six ft^3 BL/M placer gold dredge, owned by Fisher and Baumhoff. This was accomplished in 1936 near Centerville, Idaho.

Riffles were installed following the jigs in the sluices to test any losses that may have resulted, with cleanups made every two weeks. After some months when only negligable amounts of gold were found in the riffles, cleanup was extended to every two months. However, this gave the company the confidence it needed to completely remove the riffles and depend upon the jigs.

Yuba Jig Development

By the following year or 1937, Yuba Consolidated Gold Fields installed jigs in their No. 19 BL/M dredge (18 ft^3) but kept riffles in the circuit. This was followed by installations on Yuba #21 using PanAmerican jigs from the beginning of its construction.

This delay in acceptance by Yuba was reported by Charles Romanowitz (former General Sales Manager of Yuba), to be the result of an error made in their jig testing program: "Those tests involved examining the last riffle in each table sluice first, only once in a week. When the same amount of fine gold was found present on each of the riffles, this caused the belief that all the gold was being obtained. But when the test was made daily, it was discovered the same amount of fine gold was present. A better testing program then showed the real need for jigs due to the large quantities of water on the tables carrying off the fine gold in appreciable amounts."

It was in 1947, that Yuba instigated its own design effort to develop a jig that was not as tall (using less headroom), as the Pan-American. This was accomplished by moving the diaphragms to the ends of the jig instead of the bottom, saving additional headroom in the dredges.

By then, Yuba was convinced by then, of the fallacy of using riffles on dredges and in spite of strong resistance from old-time operators who clinged to the supposed security of riffles, began removing riffles from their dredges with significant results.

Tin Dredges in Malay States

Early Jig Installations
The well known dredge designer from New Zealand who moved in the 1890's to England, F. W. Payne, reported in 1926 of his observations of the use of the Harz jig on dredges mining tin in Malay states. Payne saw certain limitations in the Harz jig and suggested that the Hancock jig might work better. He also was concerned that too much was being expected of jigs and that the launders (riffles) might be better emphasized.

Perhaps his greatest contribution to early acceptance of jigs in mining for tin, was in his discussion of the need to pay more attention to classification of the ores before reaching either the jigs or launders. His emphasis was upon trying to reach some balance of size of holes in the screen plates of the trommel and advising not to remix the fines obtained at the head of the screen using smaller holes, with the coarser material recovered lower in the screen with large holes.

He referred to a classifier that was being used called "spitzkasten," but criticized it. He then proceeded to speculate on the future of jig development and recommending consideration be given to variable speed on the motor drive for fine tuning of the jigs, and being manned by personnel that could monitor them regularly.

Payne stated that only the Harz jig was being used in tin dredges on the Malay Peninsula in 1926. However, Norman Cleaveland reported that his former company, Pacific Tin Consolidated Corp., moved a BL/M dredge from Alaska to Malaya in 1922, and reported it to be the first dredge in the State of Perak to have jigs installed.

Circular Jig Development

As mentioned in History of Mineral jigs, the circular form of jig was the "first" since it was the basket jig that was the beginning of jigs. Circular sieves were common to Malayan tin sheds in their crudest forms, for many years. The Dutch State Mines developed a circular jig at their tin mines in Indonesia and were used in the shore facilities or sheds. One of those jigs was also used in Nigeria in the tin mines. Experiments had been made in other locations of Malaya on circular jigs but the first attempt to adapt them to dredges was made by Norman Cleaveland in the early 1950's, while president of Pacific Tin Consolidated Corp., in Malaya on their tin dredges.

The first such jig which he called the "Cleaveland Radial Jig," was five-foot in diameter, and had a center post-motor driven, set of radial wipping blades. These were used to maintain a uniform depth and spreading of material over the jig bed. The Radial Jig was installed on PacTin's Dredge No. 5, in mid 1950's and was followed by tests with an eight-foot jig.

Cleaveland made further developments with a 22-foot diameter jig. In early 1960, he installed a 6-foot jig on PACTIN's Dredge No. 2, and reported in 1963; " For the past eight months this jig has been handling on an average 40 yd^3/hr of (-) 5/8-inch material, with losses comparatively below those normally experienced on conventional jigs. Thus the capacity of this jig is at least 1.4 yd^3/hr/ft^2 of jig bed area, or, about three times that of the conventional jig. Deceleration of the feed as it passes over the jig bed appears to increase the capacity of a jig substantially."

Later tests with the 5-foot jig in a gravel pump mine, used a flow of 1.0" minus material fed through a grizzly, with an average feed rate of 30 yd^3/hr. This showed that the skimmers (wipping blades), could easily cope with the coarser material. (Note: Later testing made it clear, that for best jigging results, the gangue or ore-feed should stay within the size limits of (-)1/2" to (-)3/4" maximum. Subsequent tests were run on an 18-foot jig that proved successful.

Diamond Dredge Mining Applications

During the same period of the 1950's, Cleaveland was working on the solving of the problem of continuous, dredge mining of alluvial diamond deposits in Brazil. Up to that time, no one had found a way to recover diamond from the high production of a dredge as was being done with tin and gold, with a continuously operating system. Batch recovery methods were the only ones known.

Through special arrangements, Cleaveland tested his circular jig at DeBeers' Diamond Laboratory in Johannesburg, South Africa. Sub-stantial principles were established. In particular, it was discovered that the low coefficient of friction of diamond, its "frictionless property" when wetted, produced the same effect in the jigging process as heavy minerals. The result was to draw the diamonds down through the ragging into the hutch. This made possible the successful establishment of the first dredge mining operations for diamond, eventually operating five BL/M dredges in the State of Minas Gerais, Brazil (the company was Mineracao Tejucana).

Gold Dredging Applications

The Cleaveland Circular Jig was granted patents in 1967, and has been widely adapted to tin mining dredges in Malaysia, Thailand and other parts of the world since. In 1968-70, tests were run on placer gold at the Pato operations in Colombia. On one of the 14 ft^3 BL/M dredges, installing one eight-foot Circular jig alongside of the Pan-American jigs.

The tests ran for over 1.5 years, alternating from port to starboard to eliminate any conditions that might have favored one side or the other. Other variations and observations were made under several conditions, including during high sand charges, clay and jungle debris that would run through the system. The final result of the tests showed that the circular jig recovered more gold than the Pan-American jigs; was able to handle larger charges of sand and other adverse conditions, better than the Pan American jigs.

Convinced by the favorable test results, eight 9-foot diameter Cleaveland Circular Jigs were installed on Pato's Dredge No. 9, in 1974. The design was prepared by CPD for the new Colombian owners who acquired Pato at that time. The eight jigs, grouped in pairs powered by single 7-1/2hp motors, with four on each side, were elevated above the trommel.

The undersize material was pumped from a sump mounted under the trommel, to a single distributor mounted above with eight flow lines gravity fed to the primary jigs. At a later date, after using Pan-American jigs for secondary recovery, the Pan-Am jigs were replaced by four six-foot circular jigs.

A second installation of circular jigs was made on an 18 ft^3 BL/M dredge in 1980, the former Yuba #20 dredge sold to the new owners by CPD and moved to Colombia. For primary or rougher jigs they purchased three 22-foot jigs from Brazil that used the design made for the first diamond dredges, and four 6 ft. circular jigs for secondary.

IHC HOLLAND Jig Development
Pacific Tin Consolidated licensed the 1967 Cleaveland Circular Jig patent to IHC HOLLAND, The Netherlands, a dredge builder who subsequently conducted considerable tests on the jig. IHC built a number of the Cleaveland jigs for installation in dredges in Thailand and Malaysia.

Meanwhile, they began developing their own version of a circular jig (actually, a trapezoidal jig and not an infringement on Cleaveland's circular jig), removing the wiping blades that are an essential feature of the Cleaveland patent of 1967.

MK II CLEAVELAND Circular Jig

The next step in circular jig design was taken in 1979 by Norman Cleaveland. This he did in conjunction with the design development of a new, 20 ft^3 BL/M dredge for mining gold in New Zealand, as a design team member at CPD. This design development resulted in the filing of a new patent, granted in 1982, which has been termed by CPD, the licensee, as the "Mark II Cleaveland Circular Jig" (see patent at end of chapter).

There are several changes from the 1967 patent including removal of the wipping blades and its central drive mechanism. They are replaced with a central feed box with holes on the sides of the box for flow onto the bed, and four aprons that are hinged to the top of the box and slanted downwards to disperse the overflow to the outer periphery of the jig bed. This accomplishes what the wiping blades were intended to do.

The primary advantages of the new design are:
The use of the center of the jig bed, formerly blocked by the center post drive of the blades. This is considered the most efficient area of jigging and point of highest concentration; the removal of the nuisance of the drive mechanism for the blades and its maintenance; the classification function performed by the feed box, which is adjustable by altering the area of flow of the annular holes. Designs of the MK II jig are made in sizes ranging from 3 feet to 6, 9, 18 feet, or larger.

We consider along with Cleaveland, that the 9-foot diameter jig is the optimum size. The fact that the 9-foot jig uses a single diaphragm, and has a favorable lip distance relationship to feed, gives it several operating advantages (see Table A Mineral Jigs, Comparative Data, Mk II Cleaveland Circular Jigs).

Mineral Jig Alternatives

There are two basic types of jigs that are used today. One includes both rectangular and trapezoidal; in the same category, because of the flat sides or partitions, with feeding taking place at one end of each segment or channel and discharge at the other.

The other type is the circular, which has several features but common to them all is its feed in the center with the flow to discharge radiating 360 degrees over the periphery. Contemporary jigs of the latter category are of comparatively recent vintage and commercially, have two basic sources. The former type is of older design where patents have expired and may be purchased from a number of fabrication sources throughout the world.

Falacy of Riffle Systems

The earlier gold dredges, portable washing plants, and gravel pumps processing hydraulicly mined tailings, normally used Hungarian Riffle systems and various other forms. In recent years, simple washboard types of riffles using "astro-turf" material have been widely used in Alaska and Canada on small scale placer projects. However, little seems to be known among those operators of the findings and experiments that led to the adaptation of jigs to placer dredges where the limitations and high losses characteristic of riffles were discovered and proven.

One such report is quoted as follows: "The finer the gold, the more danger there is of its being carried away in the top water. The variations in the quantity of material and in the amount of water sent to the riffles frequently result in accumulations of sand, thus hindering the separation of the gold from the sand and gravel.

If the riffles are crowded beyond capacity by a sudden influx of sand, a high loss of gold during that period can be expected. Packing of the riffles by heavy constituents present in the ground, such as black sand or sulphides is another well known cause of the loss of gold. The jig either eliminates or minimizes greatly these factors contributing to the poor recovery of gold. To be saved in the riffle, the gold must settle in a swift current of water."

Arguments for Jigs

Once a particle is caught in a jig it cannot thereafter be lost, whereas it may be washed or splashed out of a riffle. A jig saves more black sand; if black sand is used as an index of efficiency, jigs far surpass riffles. Final concentrate from jigs contains no light sands; there is considerable in riffle concentrate, making a greater quantity of concentrate to clean. Riffles must be shut down for a clean-up; jigs discharge while operating.

Jigs are particularly advantageous for clean-up concentration. It is to be noted that the high ratio of concentration in roughing both on jigs and tables is bad practice generally. Improvement in placer recoveries is most likely to come by way of a treatment scheme that breaks the present roughing step on screen undersize into at least two steps.

A report by Charles Romanowitz stated that later sampling showed that the losses of riffle dredges were 35-40% of the gold but when totally replaced by jigs, recovery averaged 100% of drill values. This was true of the larger operations over the postwar years of Pato Consolidated Gold Dredging Ltd. in Colombia; Bulolo in New Guinea; Yuba Consolidated in California; and U.S. Smelting Refining & Mining Co. in Alaska; total operations collectively of over 50 dredges mining gold.

Section 5
Selection of Mineral Jigs

Primary factors
1. Total production planned for the dredge or excavation system.
2. Screen analysis of material to determine percent of production that will flow through the screen classifier to the jigs.
3. Factor for flow-rate over jig bed.
4. Experience as well as data contained in instruction manuals for use of jigs, varies greatly as to the capacity of jigs to handle feed at some ratio to the jig bed area. The character of material has a significant effect on the ability of the jig to handle the flow of material; i. e., clay, large amounts of sand, type of ore, size of feed.

Early experimentation suggested that the larger the feed grain size up to some practical limit, the higher the ratio; or, in some cases the ratio of feed volume (yd^3/hour/ft^2 area) was as high as 3:1. In practice this has usually been something closer to 2:1. In the case of lighter minerals such as cassiterite (tin), the ratio may be less than unity. However, usage indicates an average 1:1 ratio for rectangular jigs while the experience with Cleaveland Circular jigs is 1:1.5 or higher.

Example of Specifying Jigs
Dredge Production: 500 yd^3/hr to Trommel screen, 60% (-)1/2" feed; results in 300 yd^3/hr to Primary Jigs.
Jig area calculation and number of jigs:
Rectangular jigs: 300/1.0=300 ft^2 or 1-42" x 42" jig=11.0 ft^2 or 300/11=30 jigs.
Circular jigs: 300/1.5=200 ft^2 or 1-9' jig=63.6 ft^2, or 200/64=4 jigs
Secondary jigs:
Assume 10:1 reduction:
Rectangular: 300/10/Il=3 jigs (42" x 42" ea.).
Circular: 300/10/28.3=1 jig (6 ft.)(1-6 ft. jig = 28.3 ft^2)

Flow Diagrams of Placer Processing Systems

In order to make a proper design of any mine recovery system, there must be an understanding of the physical characteristics of the deposit; the associated materials, type of ore, target mineral to be recovered. Those and other details of the mine plan enable the designer to specify equipment that can process the ore and recover the mineral.

Some examples of flow sheets for dredges and portable washing plants are included in Appendix D. No representation can be made for the preference of any of the systems since they all represent varying conditions, configurations of dredges and preferences by the designers. The variations can be seen in combinations of systems that have gone into the process of developing recovery approaches and attempting to solve the problems of particular deposits.

Since the purpose of this chapter is to present the background of mineral jig development and the cases for the different types, it is left to the reader to make his own evaluations and decisions for his particular needs. A competent engineer experienced in the design of placer mining processing plants should be consulted.

Hutch Concentration

While early jig operations in lode mines of various minerals showed some with high ratios of concentration in jigs, the experience with gold and tin mining dredges have tended to be lower. Ratios of 7-15% seem to be normal and depend upon various factors but the experimentation on this aspect has not been sufficiently reported to draw any conclusions. Our company has assumed a 10% average reduction factor for design purposes.

Feed to a single 9 ft circular jig of 100 yd^3/hr, should expect to result in a flow of 10 yd^3/hr from the hutch to the cleaner or secondary jig. The example of 300 yd^3/hr to the jigs would then require that the cleaner jigs handle 30 yd^3/hr. Since there is no data that shows conclusively that the circular jig will concentrate in a higher ratio than rectangular jigs (however, some tests in Malaysia have shown a 7.0% rate for Yuba and Pan American jigs and 11.5% for circular), then a single nine-foot or six-foot circular jig would handle the secondary product; compared to three or four rectangular jigs of 42 inches by 42 inches.

Actual experience on dredges show that a conservative approach has been taken on installations of secondary jigs. Part of this may be attributed to recirculation of the concentrate designed into their system. Normally, the number of secondary jigs installed may exceed the above ratios as much as two to four times. This we believe to be more the result of inadequate results and a "playing it safe" approach.

Jig Bed Material

The jig bed is covered with a material near to the specific gravity of the heavy concentrate or mineral being recovered. This is commonly Hematite when mining casitterite or diamond, and steel shot for gold. In the case of diamond, that is the exception since it depends upon coefficient of friction rather than density.

The thickness of the bed can vary with conditions and ore that is being treated,but in gold and tin, a three-inch thickness tends to be a standard. This is common to both types of jigs. Testing with this aspect is necessary to be performed periodically by the mining engineer or metallurgist, to determine if recovery may be suffering and requires an adjustment.

The presence of clay in the feed to the jigs can cause difficulty with steel shot and frequent raking is required to loosen up the bed. Raking is essential to maintain a lively jigging bed which maintains the efficiency of the system. Some bed thicknesses may be as much as 6" which is controlled by the height of the tail board on rectangular jigs and the discharge lip on circular.

The combination of the ragging, or, artificial bed and the natural bed of feed material can sometimes build up too high and degrade the concentration efficiency and must be checked periodically.

Jig Bed Screen

The size of the ragging must be sufficiently large that it does not blind, or, stick into the screen openings. It is common to use a wire mesh screen and there are many combinations, the most widely used being one with 1/8-inch by 2-3 inches separations. This will assure however that some of the mineral will be trapped on the bed which defeats the problem of security as well as requiring more frequent cleanup of the jig bed.

The hutch product of the jig achieves the goal of less maintenance and more continuous mining, less down-time and the concentration of the mineral into a more secure, single location. In the case of gold, the concentrate can be designed to flow directly into a screened and locked room with the amalgam system.

In order to achieve a greater through-put of the target mineral, round tapered holes have been found to be more effective. Experiments have been made which show that an overall open area of 20% will produce the optimum concentration of the mineral being jigged. The 1/4-inch x 3/8-inch tapered holes are the ideal unless there are nuggets of a larger size that will be frequently encountered, in which case the holes can be larger, but not exceeding the 20% total open area.

A rubber-backed steel plate with tapered holes has been found be an effective combination and has been widely used in Malaysia. Frequent checking of the screen to insure that no blinding is occuring or to correct blinding by cleaning the screen openings, is essential to jig efficiency. Blinding may be the result however of some of the above mentioned problems; i.e. ragging size or too large an overall open space in the screen.

Section 6
Operating Considerations

Adjusting Jigs
The following basic considerations are required when adjusting jigs.
1. Stroke length of diaphragm plunger drive.
2. Speed of motor and gear reduction with resultant rpm to eccentric or plunger cycle.
3. Hutch water feed, or, gallons per minute (gpm).
4. Depth and composition of ragging, or, artificial bed material.
5. Type of screen, spacing of open area and total area exposed.
6. Primary feed to jig; grain size, clay and other obstructive content plus amount of water content.

Exposure Time of Grains
A fundamental factor relating to the concentration ratio of a jig, is the exposure time of each particle or grain to the jig bed pulsations. It is this consideration that led to the observation that to cause deceleration of the feed would have the effect of improving jigging efficiency. Thus the increasing of water flow with the feed to increase the velocity over the jig bed will have the effect of reducing efficiency.

The rectangular jig requires a certain amount of acceleration to keep the flow of material in motion. However, the circular jig has the opposite effect with the flow decelerating as it moves from the center feed to the periphery of the jig bed; thus contributing to its advantage.

Cycle Speed
The strokes per minute and the stroke length, have an influence on the exposure time of the particles to a given number of cycles of pulsion and suction which is the phenomenon that causes concentration to take place. On both types of jigs, a balance must be struck taking into consideration the flow rate over the bed with other factors such as presence of slimes, or, other inhibiting agents that will require extra exposure time to concentrate.

If the speed of the eccentric, or, plunger is increased too high either wear will defeat the process, or gravity of the particles will be exceeded that will impede rather than concentrate. For placer operations, ranges from 100 to 200 rpm or cycles/minute are normally used.

Length of stroke

This will increase the opening of the jig bed proportionate to its increase and at the same time, will increase the amount of low-grade concentrate that will be pulled into the hutch. A longer stroke also necessitates consideration be given to increasing the depth of the bed. A deeper bed can compensate partially for an increased stroke.

Length of stroke may vary from 1/4 to 1-3/4 inches, or, higher. For the rougher/primary phase of jigging it is normal to maintain a stroke length of 3/4 to 1-1/4 inches. The cleaner/secondarty phase uses a smaller stroke and in both cases varies with the combination of materials being handled, density, flow rate and presence of impeding agents such as clay.

Hutch water

This is an important factor that assures the proper functioning of the jigging process. If there is insufficient water makeup after losses to the bed resulting from the pulsing stroke, subsequent strokes will not open the bed and concentration will be degraded.

Water is admitted to the hutch through the manifold feed line and a plug cock check valve. The plug cock regulates the quantity of water admitted on the down stroke making up water lost from the hutch on the upstroke. The check valve prevents the return of the water to the line on the up stroke.

A steady head tank placed about 12 feet above the hutch will provide sufficient head for a range of flow through lines of 1/4-3/4 inches, up to 135 gpm. The metering of water to the hutch can affect the quantity of concentrate to the hutch in proportion to the water flow; the higher the flow the more suction that is exerted on the down stroke and thus higher amount of concentrate.

Tuning of Jigs

This is of paramount importance when the foregoing variations are considered. While each of the relationships are important to the concentration ratio of the jigging process, other aspects such as the presence of very fine gold, will influence the tuning of stroke, speed, ragging thickness, screen size and hutch water flow with different considerations than where only coarse gold is present.

As much advance testing with jigs as possible, using the typical materials of the deposit, is important to save checkout time when production begins. The more reliable the initial setting, the greater confidence and less down-time can be experienced. In any event, testing and sampling should be carried out after operations begin to evaluate the effects of varying the factors that influence concentration ratios and the optimization of recovery of the target mineral.

Hutch spigot

The size should be kept as small as possible, based upon a calculation of expected flow rate to prevent a buildup of concentrate and provide a continuing passage into the flow line.

Section 7

Decelerating the Flow of Material Over a Jig Bed to Increase Productive Capacity

by Norman Cleaveland

The following paper was prepared in the 1960's during his development of the circular jig.

"On conventional jig arrangements for placer mining dredges, the flow of material over the jig beds is gradually increased. This acceleration takes place because the hutch water from each jig cell is added to the flow of material. Thus, the more cells in the flow line, the greater the acceleration. That higher flow of water decreases the tendency of material to settle, particularly fine material.

On conventional jigs, water is used simultaneously as the medium in which concentration takes place and also to transport solids over the jig bed. Jig pulsations are another factor in transporting the solids but sufficient velocity of water must be maintained to discharge gangue particles in the order of (-)5/8-inch, while at the same time it is hoped that fine ore particles will settle. If the finer ore particles do not settle immediately as they pass over the jig bed, the subsequent acceleration of the flow will tend to keep them in suspension thereafter. In an effort to improve on current jig practice, two additional experimental installations have been made on Pacific Tin's Dredge No. 2.*

Firstly, conventional jigs in the secondary circuit have been arranged in a "T" formation. The feed, after passing over two conventional 42 inches by 42 inches cells in series, is split and then passed over two cells in parallel, thus changing the width of flow from 42 inches to 84 inches, with a consequent reduced rate of flow over the final cells. This arrangement also makes it possible to alter the length and speed of stroke on the final cells.

Except for the fact that this arrangement is rather awkward for limited deck space on a dredge, it seems to be an improvement and may be useful for gravel pump mines. Four such units would require 16 cells and would give a total of 28 feet of discharge lips, thus probably eliminating any need for dewatering the discharge from a conventional gravel pump handling up to 100 cuyd/hour.

Secondly, a more radical departure from the conventional is a "Cleaveland Radial Jig," patents applied for. On Dredge No. 2, this jig is 6 feet. in diameter and is fed at the center from which the feed decelerates steadily until it passes over the discharge lip of nearly 19 feet.

*This dredge first started operating in 1908 on Bonanza Creek in The Klondike. It was dismantled and shipped to Malaya in 1922 and was the first dredge in Perak to be equipped with jigs. On being moved to Selangor in 1955, the jigs were rearranged to provide additional flow lines,but the number of cells per flow line were reduced from four to three.

Presently it also has three flow lines of two cells each as well as the "T Formation" arrangement and RadialJigs. A mechanical distributor provides even distribution to therougher jigs. Skimmer blades, revolving horizontally about the center of this Radial Jig, keep the surface material moving as the water velocity drops. These skimmer blades also tend to keep the jig bed at a uniform thickness regardless of variations in hull trim.

In effect, these skimmer blades function somewhat similarly to skimmers on flotation cells but in addition cover the entire surface of the jig bed and are shaped to give the surface material a spiral path from the jig center to the discharge lip. Thus the water becomes primarily the medium in which concentration takes place and the duty of transporting the surface material over the jig bed is performed larqely by the skimmer blades.

For the past eight months this jig has been handling, on an average, 40 yd³/hour of -5/8 inch material with losses comparatively below those normally experienced on conventional jigs. Thus the capacity of this jig is at least 1.4 yd³/hour/ft² of jig bed or about three times that on the conventional dredge jig. Deceleration of the feed as it passes over the jig bed appears to increase the capacity of a jig substantially.

Similar results have been obtained from a Cleaveland Jig having a diameter of five feet. After operating on an improvised basis for several months on Pacific Tin's Dredge No. 5, this jig was subsequently moved to a gravel pump installation where it has been operating for over nine months handling material, which has passed through a grizzly with 1-inch spacings, at an average rate of 30 yd³/hour. The skimmers successfully cope with the coarser particles in the feed.

In consequence of the experience outlined above, a Cleaveland Radial Jig 18-foot in diameter, is presently in the course of being tested ashore in preparation for replacing the six flow lines of four cell rougher jigs on the starboard side of Dredge No. 5. It is estimated that this replacement jig will have a capacity of at least 250 cuyd/hour of -1 inch material. If such figures are confirmed in full scale operations a number of advantages become apparent, such as:

1. Two such jigs, one on each side of the dredge, would substantially increase present jigging capacity and this increased capacitycould be utilized by increasing the size of perforations in thescreen plates of the main trommel which in turn would then be able to handle more material from the bucket line. At present the bucketline is dumping only 21 buckets-per-minute, therefore an increase of at least 25% in overall capacity of the dredge seems a reasonable possibility.

2. With only two jigs on the dredge, a simple splitter under the main trommel should solve distribution problems.

3. There would be far fewer working parts and less power, labor and deck space required.

4. The skimmer blades give excellent control over the surface contour of the jig bed and also control the rate of flow of material. Thus, by increasing the speed of the skimmers, greater capacities than those presently estimated may be possible.

5. At the speeds currently being used these skimmers provide a practical means of spudding the bed mechanically.

While variations in hull trim have presented no problems for the small jigs, it is expected that some adjustment will be necessay for the larger jig and a simple means of doing so has been provided. With limited deck space, multiple radial jigs of the smaller diameters seem to have the following advantages.

1. Rate of deceleration is higher.
2. Length of discharge lip relatively longer.
3. Probably less weight and first cost.

A single large radial jig may be superior on the followingpoints:

1. Capacity greater for deck space available.
2. Simplified distribution.
3. Fewer moving parts.

A jig of 18-foot in diameter, is the largest that can be conveniently installed on Dredge 5. Operating experience with this 18-foot diameter jig should provide valuable data to guide future designs. Experience acquired to date with smaller radial jigs suggests that a substantial increase in the productive capacity of most tin dredges is possible.

General Comment

There is nothing new about the idea of a Circular Jig. In fact, the original jigs of antiquity were probably baskets, therefore probably circular. Primitive jigs, some circular, are still used extensively. The circular sieves common to Malayan TinSheds are a form of jig and like most primitive jigs are used on a batch basis and are skimmed by hand.

Buddle jigs were amongst the earlier mechanical jigs whichoperated on a continuous rather than a batch basis. These jigs are circular and are still widely used. The Dutch State Minesdeveloped a circular jig which is presently operating in Indonesia. At least one of these jigs has operated in Nigeria. In Malaya the Foo Nyit Tse mines at Kampar have experimented with a circular jigand others may have also.Handbooks on mining and ore dressing show several forms ofcircular jigs. Such handbooks also refer to various forms of theHancock Jig, a type of jig in which the rectangular bed is mounted in a mechanically actuated basket which carries the feed over the bed with a series of grasshopper like jumps. Thus water plays a minor role in transporting the solids and becomes primarily the medium in which concentration takes place.

Mr. A.D. Hughes, a Director of Pacific Tin with long experience in Malaya, installed a skimming device on conventional Harz type jigs operating at low capacity for the recovery of diamonds in Brazil during 1928. Mr. Hughes has been active in some of the expermental work Pacific Tin has undertaken on jigs during the past ten years, particularly the work done in co-operation with the Natomas Co., one of California's oldest gold dredging firms. Mr. John Munslow Chief Engineer of Pacific Tin, has been responsible for preparing the details of design for the Radial Jigs and has otherwise played an active part in the company's experimental work. In preparation of this paper the writer is indebted to the following persons: C.G. Carlson; A.D. Hughes; F.S. Miller; C.M. Romanowitz; E. von Bolhar; W.M. Warren; F.A. Williams(Kuala Lumpur, Malaya; 3/13/63).

A—Workman carrying broken rock in a barrow. B—First chute. C—First box. D—Its handles. E—Its bales. F—Rope. G—Beam. H—Post. I—Second chute. K—Second box. L—Third chute. M—Third box. N—First table. O—First sieve. P—First tub. Q—Second table. R—Second sieve. S—Second tub. T—Third table. V—Third sieve. X—Third tub. Y—Plugs.

Fig.1-Sluicing gold and jigging with circular sieves, c.1556 (reprinted from "De Re Metallica", pg. 291)

FIGURE 1-SLUICING GOLD AND JIGGING WITH CIRCULAR SIEVES, C.1556 REPRINTED FROM "DE RE METALLICA," PG. 291).

MINERAL JIGS COMPARISON

A.	B.	C.	D.	E.	
SIZE DIA FT	AREA[1.] SQFT	HUTCHES NO.	FLOW[2] CY/HR	LIP FT	RATIO[3] D/E
18	254.3	4	381.5	56.5	6.8
9	63.6	1	95.0	28.3	3.4
6	28.3	1	42.4	18.8	2.3
3	7.1	1	10.6	9.4	1.1

COMPARISON WITH YUBA/PAN AMERICAN JIGS
(Rectangular)

4 Cell	49.0	4	49.0[4]	7.0	7.0

1. Area of jig bed exposed to jigging.

2. Assumes flow ratio of 1.5 for circular jigs. Varies with composition of ore; the coarser the higher the ratio, within limits.

3. Since concentration increases with deceleration, a lower ratio is an improvement.

4. Rectangular jigs have added friction with side-walls thus higher flow rates with shorter lip distance, thus reducing capacity to 1:1 or less.

TABLE A -MINERAL JIGS, COMPARATIVE DATA, MK II CLEAVELAND CIRCULAR JIGS

United States Patent [19]

Cleaveland

[11]   **4,310,413**

[45]   **Jan. 12, 1982**

[54]   **CIRCULAR JIG**

[76]   Inventor: **Norman Cleaveland,** P.O. Box 4638, Santa Fe, N. Mex. 87502

[21]   Appl. No.:   **160,081**

[22]   Filed: **Jun. 16, 1980**

[51]   Int. Cl.³ **B03B 5/12; B03B 11/00**
[52]   U.S. Cl. **209/456; 209/498**
[58]   Field of Search 209/455, 456, 459, 460, 209/475, 476, 488–489, 497–499

[56]   **References Cited**

U.S. PATENT DOCUMENTS

431,607	7/1890	Monteverde	209/498
1,100,971	6/1914	Hambric	209/489
2,287,748	6/1942	Pardee	209/456
3,273,714	9/1966	Cleaveland	209/498 X

Primary Examiner—William A. Cuchlinski, Jr.
Attorney, Agent, or Firm—Edward B. Gregg

[57]   **ABSTRACT**

Circular jig for gravity separation of minerals, e.g. gold and tin ore, comprising a screen, a grid forming cells over the screen, a central jig box of polygonal, e.g. square shape, pulsating means to cause the jig bed to pulsate and aprons attached to the side walls of the jig box extending outwardly from the jig box to distribute overflow from the jig box to portions of the jig bed remote from the jig box.

7 Claims, 3 Drawing Figures

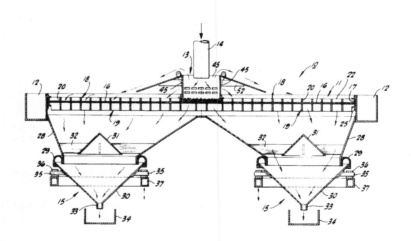

UNITED STATES PATENT NO. 4,310,413 CLEAVELAND CIRCULAR JIG

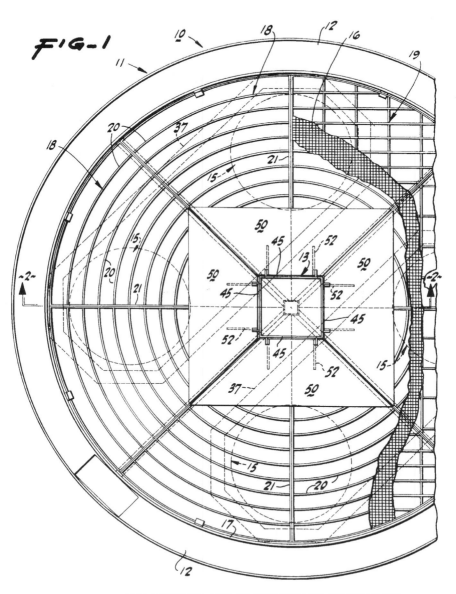

FIG-1

UNITED STATES PATENT NO. 4,310,413 CLEAVELAND CIRCULAR JIG

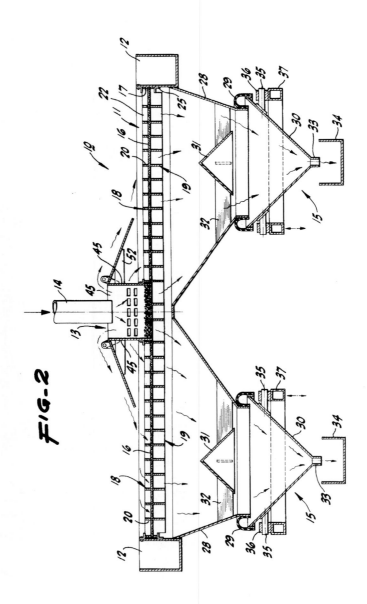

FIG-2

UNITED STATES PATENT NO. 4,310,413 CLEAVELAND CIRCULAR JIG

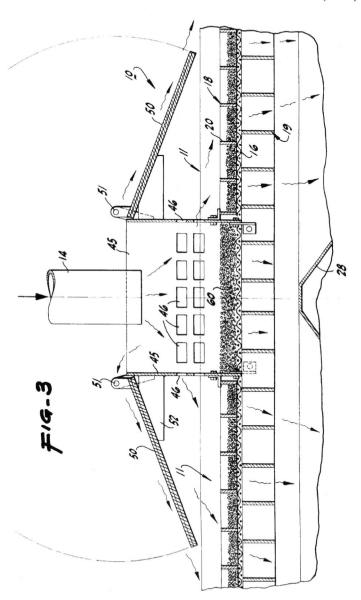

FIG-3

UNITED STATES PATENT NO. 4,310,413 CLEAVELAND CIRCULAR JIG

1

4,310,413

2

CIRCULAR JIG

INTRODUCTION

This invention relates to mineral dressing in general and more particularly to circular jigs used to recover or upgrade ores by gravity concentration which results from a pulsating action on the material being treated. This material consists of sized solids being transported by water and generally known as the feed or the pulp which is subjected to pulsation as it passes over the jig bed. This bed is supported by a screen above a water-filled hutch which is fitted with a rubber diaphragm. The pulsations on the jig bed occur when the diaphragm is actuated by an eccentric. The screen can be of woven wire or of light steel or rubber plates perforated with tapered holes.

Normally jig beds are square and the pulp passes over two or more jig cells in series. This results in a slight acceleration in the pulp flow because of the additional water added to each hutch to provide a very mild counter flow upward through the jig bed. Acceleration of the pulp flow, however slight, is detrimental to the recovery of heavy minerals. Also the jigging action is less effective along the longitudinal sides of the jig bed because of the complications due to friction, known as "side wall effect."

It has long been recognized that, ideally, a jig bed should be circular and fed at the center. On such a jig bed, the flow of pulp will decelerate as it flows towards the periphery. Deceleration of flow contributes to the concentration of particles with high specific gravity. Furthermore, a circular jig needs no side walls and therefore does not suffer from their effect.

The flaw in the circular jig concept becomes apparent under heavy loads. Then the pulp tends to pile up or "pyramid" at the center and thus dampen pulsations where in fact they should be strongest. This dampening at the center causes the pulsations to be stronger than desirable near the periphery where excessive pulsations are detrimental. To overcome this detrimental feature, U.S. Pat. No. 3,273,714 was issued for skimming devices that contour the jig bed surface to keep the depth of bed nearly constant and thus assure that the jig's pulsating effect is generally uniform over the entire jig bed. But such skimmers also have drawbacks because the essential mechanism not only blanks off the most effective concentrating area but also consists of moving parts that require power and maintenance.

The device presented in this application improves the performance of circular jigs by making it possible to take full advantage of the normal pulsating action common to all jigs but in addition to obtain the concentrating influence of both the deceleration of the pulp flow and the boiling action around the feed intake, this without moving or power consuming devices such as skimmers.

One embodiment of the invention is shown by way of example in the accompanying drawings, in which:

FIG. 1 is a top plan view of the jig partly broken away to reveal interior construction;

FIG. 2 is a section along the line 2—2 of FIG. 1; and

FIG. 3 is a fragmentary vertical sectional view of the jig drawn on a larger scale than that of FIGS. 1 and 2 and showing certain details of construction.

The drawings included herewith outline a jig with a bed of some 16 feet to 18 feet in diameter and with four pulsating diaphragms. For jigs with diameters from 6

feet to 9 feet, a single diaphragm would be adequate. Also for larger jigs, say from 20 to 25 feet or more in diameter, eight or more diaphragms may be required. Although dimensions may vary, the principles involved do not.

Referring now to the drawings, the jig is generally designated by the reference numeral 10 and it comprises a jig basket 11, a tailing sluice 12, a jig feed box 13, a feed inlet 14, and pulsating elements 15 of which there are four, one for each quadrant of the jig basket. The jig basket is constructed as follows:

A segmented circular screen or perforated plate 16 extends from the center of the jig to the inner wall 17 of the sluice 12 and is associated with a grid structure comprising an upper grid 18 and a lower grid 19. The upper grid 18 is formed by arcuate concentric segments 20 fixed at their ends to radial members 21. This grid forms cells 22. The lower grid 19 serves to support the screen 16. The screen and grid structure is attached to the walls 45 of the jig box 13 and to the inner wall 17 of sluice 12 by brackets 25. The outer brackets are attached to the wall 17 by wedges to allow ready removal.

Each pulsating element 15 comprises a hutch 28 which tapers downwardly and is connected at its lower end to a flexible diaphragm 29 and empties into a cone 30. An inverted diffusion cone 31 is affixed by brackets 32 to hutch 28. Cone 30 empties through a spigot 33 into a sluice 34 which may be circular and has a suitable outlet (not shown). The attachment of the cone 30 to the hutch 28 is by way of the diaphragm 29. The cone is supported by trunnions 35 rotatable in bearings 36 which are supported by a rocker beam 37. As will be seen from FIG. 1, each rocker beam 37 operates two of the pulsating mechanisms 15 and it is caused to rock by suitable means (not shown) so that as one of each pair of cones goes upwardly the other one goes downwardly. This rocking imparts a pulsating motion to the quadrants of the jig basket.

The jig feed box 13 is square in cross section and its side walls 45 are formed with openings 46 and it is open at the bottom to the central portion of the screen 16. Four aprons 50 are hinged at 51 to the four side walls 45 of the jig fee box and are supported in inclined position by brackets 52 which are mounted on the side walls of the jig box. The aprons are shown as having square corners but the corners may be rounded. The aprons may be plain or perforated. The jig feed box 13 may be polygonal, e.g. square as shown, or it may be round provided the upper edge is modified to accommodate the aprons 50.

As shown in FIG. 3, steel shot or punchings 60, called "ragging," are deposited on the bottom of the jig feed box. The ragging in the annular spaces of the upper grid 18 is usually particles of hematite or material of similar specific gravity.

In operation, pulp is supplied to the feed box 13 through inlet 14 at a rate such that it overflows the feed box and is directed radially outwardly by the aprons 50 into the quadrants of the upper grid 18. The pulsating motion imparted to the jig agitates the ragging and other loose material in the jig bed. Another portion of the pulp flows out of the jig box through openings 46. The pulp flows generally radially outwardly and carries the lighter minerals to the tailing sluice 12. The heavy minerals are retained in the jig bed. All but oversize particles of the retained minerals pass through the ta-

4,310,413

3

pered apertures of screen **16** and into hutches **28** en route to sluice **34** for further treatment.

The perforations in the screen **16** and in the aprons **50** are advantageously tapered with the narrow ends of the taper uppermost. The diameter of the narrow upper ends of these perforations determines the size of material that passes through the screen and the inverted taper prevents clogging.

Among the advantages of the apparatus thus described and illustrated are the fact that the entire area of the sluice feed box is available for separation of heavy minerals. It is in this area of intense agitation that the major part of the concentration occurs.

The aprons help prevent "pyramiding" of pulp that hinders normal pulsations over the entire jig bed and at the same time promotes the uniform outward flow of the pulp's lighter fractions. Under some conditions, perforations in portions of the aprons improve the distribution of pulp on the jig bed. From time to time the hinged aprons may be lifted for access to the inner portions of the jig screens or grids.

Should the jig be mounted on a vessel subject to wave action, the clearance between the outer edge of the apron and the jig bed may be adjusted to choke temporarily the resulting flow surges, and thus help keep uniform the time during which the feed is subjected to both the jig's pulsations and to the boiling action around the feed box.

It will therefore be apparent that novel and useful apparatus and method have been provided for separating heavy minerals from light minerals at rates of about one cubic yard per hour of material per square foot of jig bed.

I claim:

1. Apparatus for gravity separation of minerals comprising:

4

(a) a circular jig basket having an outer wall, a screen extending from the center of the basket to the outer wall and a grid structure superimposed on the screen into which heavy minerals may sink and pass through the screen, such grid structure and screen providing a jig bed,

(b) a jig feed box located centrally on the screen, the walls of such feed box having side openings above the screen through which pulp may flow onto the jig bed,

(c) aprons attached to the upper part of the jig feed box, extending radially outwardly therefrom and widening outwardly therefrom whereby the several aprons form a continuous flow surface over the part of the jig bed neighboring the feed box to direct the overflow from the feed box outwardly above the jig bed and then onto portions of the jig bed more remote from the feed box, and

(d) means for imparting pulsating movement to the jig bed to agitate pulp delivered thereto.

2. The apparatus of claim **1** wherein the jig box has a polygonal shape and there is an apron attached to the upper edge of each wall formed by a side of the jig feed box.

3. The apparatus of claim **2** wherein the feed box is square in horizontal cross section and there is an apron hinged to each side wall.

4. The apparatus of claim **1** wherein the aprons are hinged to allow access to the screen and grid beneath them.

5. The apparatus of claim **1** wherein the aprons are imperforate.

6. The apparatus of claim **1** wherein the aprons are perforated.

7. The apparatus of claim **1** wherein the aprons are adjustable as to the clearance between the outer edge of the aprons and the jig bed.

* * * * *

CHAPTER V
PLACER MINING OPERATIONS

In the foregoing chapters, the process of finding, evaluating and designing methods of mining placer deposits has been discussed and defined. The types of equipment available and used have been presented. Once all those steps have been followed and equipment mobilized, the operation of the equipment becomes the next essential consideration.

Each step of a mining plan is important to that ultimate result; return on investment, success or failure. The best minds can be employed in the initial stages into mobilization of the equipment, only to be confounded with an improperly managed operation; or unforeseen problems that produce the same result.

This chapter will attempt to point out some of the learned lessons in placer mining operations but assuming that the decision process has resulted in a proper selection of equipment that has a chance for profitability. This would include a valid process of obtaining "proven reserves" that will support the chosen mining operation.

Section 1
Mining Reserve Calculations

Regardless of the equipment selected, the proven reserves need to be plotted on a map or chart that can be used to determine the most efficient and profitable course of mining. Using the tenor value calculations from each drill, shaft or bulk sampling location a plot is made of the complete area. There are two common geometrical methods used to compute volumes of influence in placer reserves: (1) Ploygon, and (2) Triangular. Calculations to determine the accuracy of the two methods compared with each other, have been made in the past with no significant difference found. The polygon method lends itself better to computer calculations while the triangular is often used in the field because it is an easy method to handle with drafting tools and simple calculator.

The purpose of the calculations is to obtain an average value for a given ground area and its volumetric reference to the cutoff point of the drill or shaft hole. We consider the polygon method to be more accurate when a computer can be used for area calculations, than the triangular method. Given a plot of drill/sample holes and their values the following is an example of the computation. The area of each polygon is made by the influence of the hole in an area to the distance and direction of adjacent holes. The purpose is to find an area that is influenced by that hole determing boundary lines from the nearest holes in each direction.

A = Area of ploygon-ft^2 (determined either by planimeter or computer).
V = Volume-yd^3 (area X cutoff depth-ft/27)
T = Tenor of sample/drill hole-mg/yd^3 (computed to cutoff depth) x
V = Tenor of polygon or mg T/31.1 mg/Troy oz =Troy oz estimated in polygon volume

From this the reserves can be totalled from all the polygons. The same steps can be substituted using the triangular method (see Figure 1-Polygon Map, plotting reserves of a large gold placer for dredging).

Section 2
Mining Course Design

A large volume reserve requiring a dredge or dredges, must be mined taking into consideration a number of factors. Most placer minerals whether gold, platinum, diamond or tin, have low tenors in terms of volume to be dredged. This means that all sampled mineral-bearing material must be handled and processed.

When there is overburden, the decision has to be made of whether to dredge it all with a single dredge, or strip by other equipment. Factors such as the swing of the dredge, slope of the material to the digging face, stepping forward and material not dredged as a result of the geometry of the cut, cleaning of bedrock or the cutoff point; all enter into the decisions of what course to plot.

An example of a large scale, gold placer dredging operation with parallel dredge courses can be seen in Figure P23.0 The two BL/M 14 ft^3 dredges are mining the river flats on both sides of the river (Nechi), Colombia. In this case, the dredges are mining with the flow of the river, downstream.

Whether to mine one way or the other requires knowledge not only of the deposit, the behavior of the river during seasons, but a long-range view of the mining program. On small deposits, the high grade location usually determines where the mining starts. Many factors need to be considered but often are influenced by the shortest time to return the capital investment.

Dredge course decisions and calculations will vary with the type of mining equipment used. The design of the course should therefore have the involvement of operational personnel who know the behavior of the equipment in order to compensate and optimize recovery in as few movements or retracing steps, as possible.

Discussions in the foregoing chapters cover several aspects that influence dredge mining courses or approaches. (Some of the best examples are contained in Appendix A, B, C-Case Studies of Bulolo, Nechi and Choco operations).

Section 3
Personnel for Placer Mining

The problem of acquiring competent personnel to carry out a successful placer mining project becomes more critical with each passing year. With few commercial scale, successful placer mining operations left the well of trained personnel becomes smaller. Often working in remote and hostile areas, it takes special individuals to provide their skills and endure the privation and severe working conditions characteristic of those regions.

The trend in overseas operations has been to rely more upon local personnel, setting up training programs to school indigenous laborers. It is essential however that a nucleus of expatriate personnel be in charge for at least the first few years, with communications to headquarters as reliable as possible. Minimizing this factor for whatever reason, has resulted in either high losses of production or the failure of many operations.

Project Failures
In the past several years there have been an alarming number of failed, placer mining projects. These are projects for the most part that are in remote areas that have not been mined before with modern equipment. Critiquing them has shown a combination of inadequate sampling and evaluation, plus improper selection of mining equipment and operating procedures. But underlying all of this appears to be a lack of involvement of experienced placer mining personnel in the entire process.

What can be done when there are few professional, placer mining engineers or geologists remaining in the world, is a difficult question to answer. Unfortunately, there are many professional engineers and geologists who profess to know or assume that placers are too simple for them not to know, how to solve the problems.

They then proceed to organize and promote placer projects for investors and major mining companies. Combine this problem with members of management who become wedded to a project decision; reluctant to admit the mistakes when apparent, in time to shut down the project or to seek out experienced engineers to redirect the effort.

Litmus Test

Evaluation of key personnel for a placer mining project should be predicated upon experience and knowledge with successful, commercial scale operations. The next stage of litmus testing is that there must at least be someone in the planning and review stage, who is an adherent with proven principles of placer mining at all stages.

Our experience has been that mining engineers with placer experience are an important key to success. Since a placer deposit is a matter of geomorphology and not hardrock geology, there is a tendancy for geologists to confuse the two and creates an operation that is too costly to be borne by the low average tenor, typical of placers.

Organization

As an example of what form of organization is needed for a large-scale placer dredging operation, Figure 1 Organization Chart shows a subdivision of personnel and functions. This will vary with the number and size of the dredges but is representative of what functions are needed.

Section 4
Essentials of Operation

Discussions in this book illustrate the choices to be made in selecting the most efficient equipment for placer mining. Factors such as the deposit size and characteristics should be the primary governing influence on what equipment is to be used. Often however, it is a matter of capital cost and personal choice that introduces improper methods and equipment. Decision processes in making a feasibility study on a proven reserve, should follow the steps outlined in Chapter II Feasibility Analysis. Some of the aspects to plan and implement are applicable in varying degrees to all placer mining operations regardless of equipment used but are mainly a matter of scale. Some important steps are summarized as follows.

1. Establish an operational plan based upon crew time; two or three shifts. Procedures and supervision of what happens in the middle of the night to cope with breakdowns and emergencies; production guidelines, repair and maintenance of critical functions must be designed, set out in writing and monitored.

2. Regular repair and maintenance shutdown needs to be planned, such as the first two hours each morning and one day-per-month, providing key personnel for those times. Outside supplier personnel need to be scheduled well in advance for their part in preventive maintenance. Principal suppliers and manufacturers are usually equipped to provide regular maintenance procedures to follow for their particular equipment.

3. Marketing of the mineral and by-products should be handled separately but well in advance to prevent excess stockpiling or decisions after the fact that may lead to reduced prices. Shipment of gold bars or other valuable minerals, should be carefully scheduled to minimize security problems.

4. Cost effective analysis of operations should be a continuing process. If early decisions were made that result in rehandling of material, high cost of operation, pilferage of the mineral, attempts should be promptly made to rectify the problems. Delays or indecision in these key matters has been the ruin of many mining projects.

Section 5
Operational Examples

The following brief discussions will relate to Figures in the Illustration Section, showing actual mining operations in various areas of the world. No two placer mining operations are alike, and therefore it is sometimes best to relate to projects where particular problems and conditions were met and solved.

Gold Dredging in Colombia
Figures P10.1, P13.1, P23.0, illustrate large scale, gold placer operations under Pato Consolidated Gold Dredging Ltd., operating from 1932-74 on the Nechi River, Antioquia, Colombia, South America. (Subsequently sold to Colombian interests and still partially operating).

The dredges shown are part of five standarized, 14 ft^3 BL/M dredges achieving the highest recorded average production for that size of dredge; up to 550,000 yd^3/month. Bucket speed averaged 30-32 bpm, digging to 100 ft in some areas.

A large support facility with complete shops, foundry and housing was constructed along with an airport, modern camp for personnel, their families and management. Standardization of dredge size contributed greatly to high efficiency and low cost operations. They established a hydro-electric plant up river and provided all electric power for their dredges and camp facilities. Operating costs were as low as $0.17/yd³ in the early 1970's, because of those efficiencies. It is estimated that over 3.0 million Toz of gold and 1.0 billion yd³ of material were produced during the 42 years of operation.

Alaska Gold Dredging
In Figure P10.2, a BL/M 9 ft³ dredge owned by Alaska Gold Co., is shown mining gold in Nome. Except for the streams, all reserves are in permafrost and have to be thawed prior to dredging. This is mostly done by drilling holes during winter to bedrock; an one-inch pipe dropped inside the holes and ambient temperature water continuously pumped into them during summer months, until thawed.

Operations may be from May to October or longer, depending upon weather and temperatures. Repair and maintenance is carried out primarily during winter months, then operations and materials pushed to the limit during summer availability. Tenors of the deposit approximate 300 mg/yd³.

Brazil Tin Dredging
In Figures P11.1-.2, a floating wash plant fed by backhoe from the adjacent land, is used to mine and process cassiterite/tin placers in the Brazilian jungles. There are numerous backhoe/wash plant operations mining tin throughout the Amazon basin in Brazil. Figure P19.0 shows a BWS/M dredge feeding a separate wash plant, also in Brazil. The higher-volume capability of the BWS/M dredge has become popular for tin mining in Brazil and has contributed to the record production levels of tin by Paranapenema.

Brazil Diamond Dredging
Figure P24.0 shows a BL/M dredge mining diamond placers, in Minas Gerais, Brazil owned by Mineracao Tejucana. At one point they had a fleet of five BL/M dredges, equipped with Cleaveland Circular Jigs (22 ft. diameter on primary). Some gold occured with the diamond which was also recovered in the circuit. The diamond tenor was very low requiring high volume production to make it pay. It was on this

project during the exploration and development in early 1960's, that Norman Cleaveland designed and tested his Circular Jig leading to the first patent in 1967.

New Guinea Gold Dredging-Bulolo
In Figure P12.1, one of the Bulolo BL/M dredges is shown that was constructed through the predecessor company of CPD. The components were shipped by ocean cargo vessel and off-loaded into long boats and carried to the beach near Lae, and air-shipped in German Junker tri-motors (circa 1931-37) to Bulolo. This was the first recorded, air freight lift and has been the subject in recent years of a movie documentary celebrating the history of flight.

A total of eight BL/M dredges were built and installed in the project; one 6 ft^3 and seven 10-1/2 ft^3(see Appendix Case Study-Bulolo). One of the most successful and profitable, gold placer operations paying stockholder/investors substantial returns.

This project was developed under Placer Management Ltd., using technical direction of CPD's predecessor, Placer Management Ltd., while simultaneously developing and operating Pato Consolidated Gold Dredging cited above, in Colombia. A total of 20 dredges were in operation at one time between the two companies, producing gold.

Peru Gold Placer Dredging
Figure P13.2 shows the BL/M 9 ft^3 dredge, mining gold placers at 17,000 ft elevation in the Andes. Originally installed and owned by Natomas Co. it operated from 1962-71. Modified from a small hull and superstructure it was top heavy and capsized two or three times, before being shut down. Heavy clay combined with angular/moraine cobbles made mining difficult, in addition to the high-altitude and its effect on personnel. The dredge was dismantled and transported from Northern California by train to a ship, then to Lima, Peru. From there it was taken by train into the Andes to about 10,000-foot elevation. It was then trucked from the railhead about 150 km to the mine site and

erected. Water was impounded behind a shallow dam in earlier years for hydraulicking, provided from scant rain and runoff from nearby glaciers. This was a constant problem of sufficient water and stands as a deterrant to future development of the very large gold placer deposit.

Malaysia Tin Dredging

Figure P14.1 shows a large BL/M 24 ft^3 dredge(F.W. Payne design), mining tin placers in an inland pond. The dredges at one time numbering over 65 in Malaysia, mine into the jungles after timber is cleared. The stern gantry suspends a long sluice and conveyor belt to distribute the fines which are mainly in clays and spread out broadly. Slime buildup is a major problem that causes recovery problems in the mineral jigs as well as at the digging face.

Bolivia Placer Dredging

Figure P12.2 shows a BL/M 11 ft^3 dredge mining gold in Bolivia, owned by South American Placers/COMSUR; originally installed by CPD for International Mining Corp.

Figure P14.2 is a BL/M 14 ft^3 dredge mining tin. Owned by ESTALSA AG, operated successfully for several years together with floating wash plant. A Yuba dredge it was moved from California to Bolivia by CPD.

Figure P25.2 is a BL/M 7 ft^3 dredge (ADL/PAYNE design), mining tin under very difficult conditions; a narrow canyon with high cliffs and subject to flooding when storms hit the higher mountains. It was erected new on site by local personnel of COMSUR, and later salvaged after capsizing, the hull and treatment plant re-designed by CPD and returned to operation.

Indonesia Offshore Tin Dredging

Figure P15.1 is a BL/M 22 ft^3 dredge (PAYNE design), operating offshore of Banka Island. Built in Indonesia by IHC/McDermott, it was towed to the site for operation.

Holland Sand Dredging

Figure P15.2 is a deep mining system, S/M dredge using a submerged pump to 250 feet to reach sand deposits in Holland. The pump is a positive displacement system that is not limited by depth.

Rutile Dredging

Figure P16.1, a CS/M dredge mining beach sands for rutile in Australia, with trailing processing plant including spiral concentrators.
Figure P16.2, a CS/M dredge mining rutile in Florida.
Figure P20.1, a BWS/M dredge mining rutile in Western Australia.
Figure P20.2, a CS/M dredge also in Australia rigged for deep digging.

English Channel S&G Dredging

Figures P17.1-.2, TH/M seagoing, self-propelled hopper dredges mining and processing of sand and gravel deposits in the English Channel. Large production results from a fleet of over 100 vessels and several company ownerships; providing construction S&G for UK and European consumers.

California Gold Placer Dredging

Figure P18.1-.2, is the Yuba #21 BL/M 20 ft^3 dredge mining near the Yuba river, northern California. The last of the Yuba operations, cited earlier, of 22 total BL/M dredges mining from 1903. Over 1.0 billion yd^3 of material, dredging in successively deeper clay lenses in three to four passes over the some 64 years, and finally now to bedrock approximately 200 ft below surface. Figure P26.1 is the largest Yuba dredge built, BL/M 18 ft^3 with digging depth of 124 feet. Mined gold placers for about 30 years near Yuba River, shut down and in 1974, moved to Nechi River, Colombia where it is now mining gold with four other BL/M dredges.

Dead Sea Salt Dredging
Figure P21 is a Dustpan-type of suction dredge, specially designed and built by R.A. HANSON CO., for the Jordan side of the Dead Sea, mining various chemical salts in shallow depths.

Small-Scale, S/M Gold Mining
Figures P22.1-.2 are examples of suction dredges attempting to mine small gold placer deposits on a random basis. Bedrock, proven reserves, mine plans or dredge courses are not considered in this type of mining, nor are they applicable.

New Zealand Gold Dredging
Figure P25.1 is a BL/M 20 ft^3 Yuba dredge mining on the Taramakau River, South Island of New Zealand. Built in 1939, it operated until 1982, digging in cobbles and boulders to 85-foot maximum depth. Built extra heavy structurally to withstand the difficult digging.

Ghana Gold Dredging
Figure P26.2 is one of the BL/M 6 ft^3 dredges of Dunkwa Gold Fields, Ghana, Africa; mining gold placers for many years. Difficult mining with sticky clay and deteriorated infrastructure, compounds the problem of profitable mining but it continues as an unsolved but high potential, gold placer deposit.

Conclusions
Modern Placer Mining is still an alternative to mine selections and should be pursued as part of every large mining company's exploration and development program. Given the problem of scarcity of experienced placer mining engineers, this should not deter the choice of placer mines but rather, to give emphasis to finding and evaluating those who can properly assess and develop such mines.

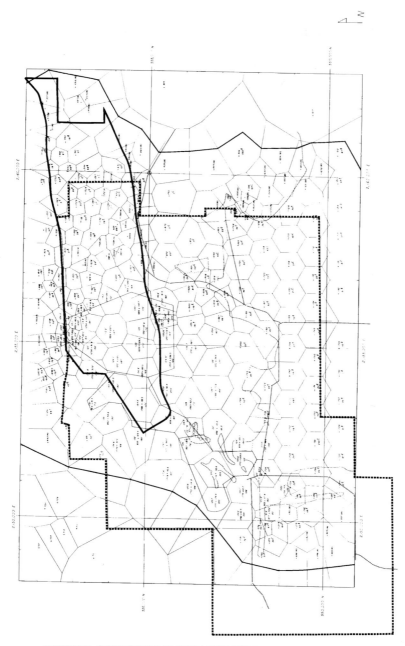

FIGURE 1 -POLYGON MAP OF LARGE GOLD PLACER RESERVES

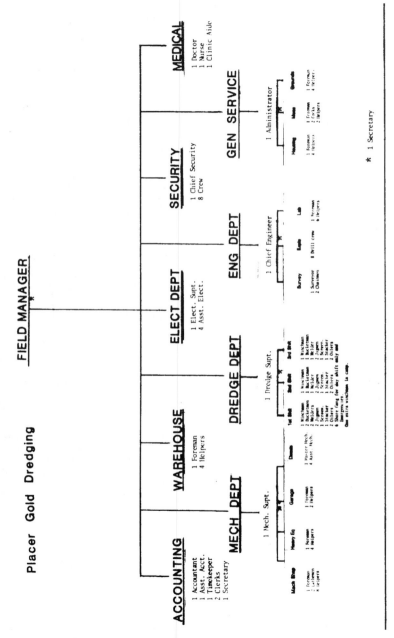

FIGURE 2 -ORGANIZATIONAL CHART, PLACER GOLD DREDGING

APPENDIX

CASE STUDIES

Appendix A
Bulolo Gold Placer Development Project, New Guinea

Early Beginning
First indications recorded of gold being in New Guinea dates to the 1500's, from reports of Italian and Spanish explorers. Although it wasn't until the late 1800's when serious exploration activity began to occur by British and German explorers. Soon gold production began and in the late 1800's, constituted over 54% of the total value of the colony's exports. However, it was not until the 1920's that the richness of the placers in the Bulolo region was established.

Exploration of Bulolo
In the late 1920's several groups from Australia, Britain and the United States, made serious exploration journeys into the Waria River. However, conflicts between the German and Australian possessions with the prospects near the border caused a further deterrent to development.

In circa 1928 the Yuba Consolidated Gold Fields group evaluated the project and turned it down because of their inability to solve the logistics problems. The question they asked was; "how do you move dredges over a mountain range and into the jungles that are almost impenetrable, with jungle fevers, unfriendly natives, and no roads?"

One prospector, Mr. Levien, persisted in lining up claims during the 1920's, observing and doing hand mining and sampling. He continued to accumulate claims to add to the potential volume, while trying to line up financing to develop the prospect.

A group of six other prospectors conducted further exploration in the Bulolo region and began to get an idea of the dimensions of the prospect. Figure 1 Map of Goldfields, District of Morobe, identifies the general location of the prospect in the center of the map.

The first phase of the project started with "Guinea Gold No-Liability Co.," registered in 1926. Centered in Edie Creek, an option was taken on part of Levein's property on the Bulolo flats and some pitting was conducted for sampling purposes, reaching what turned out later to be a false bedrock at about 15 feet. However, there was little interest in deeper gravels at that time, because no one could envision how dredges could be brought into the area anyway, therefore contemplated only light machinery and simple mining methods.

By 1907, hundreds of hand miners that had rushed to the Bulolo area had picked the eyes out of the rich shallow deposits and it was left to dredging to exploit the deeper reserves. From 1926 to 1929, the Guinea Gold Company set up an airport in Bulolo and an air service from the coast, which began the germ of the idea and led to the buying of three new German Junker aircraft (G-31) trimotors, to begin the largest cargo airlift attempted to date.

Exploration Methods
In 1927 sampling was done by sinking lines of shafts. This was because it was thought to be impossible to bring in a power drill. Portable hand drills were not used because of the frequency of boulders in the gravels, which they thought would make drilling difficult.

Samples taken at three-foot vertical intervals, indicated high-values in an estimated 5.0 million cuyd's in one area and 400,000 cuyd's in another. Bedrock was thought to occur between 12 and 18 feet, though only six shafts bottomed. It was then thought that a small dredge could handle the deposit profitably.

Testing continued into 1929 with the same misconception of a false bedrock, a common characteristic of placer deposits. Several years later it was found that the real bedrock approached 200 feet in some places. This mistake led to the construction of the first two dredges for shallow dredging.

Eventually, this was an advantage as in the later years the deposits played out. However, it was probably a bad mistake in view of what production could have been obtained with those two dredges if they had been deeper digging and had larger buckets.

During those early stages of development, the promoters attempted to raise funds on the public market to develop prospects. The fact that an airline had been built and was operating, carrying people and supplies into the jungle areas, provided the cash necessary for the company to get going.

It was at this point that Placer Development Ltd. of Canada came onto the scene and secured options and terms that would make the project go into a development phase successfully. Placer was at that time, a small mineral exploration company from Vancouver, BC (Canada), under the direction of an Australian financier named Freeman and a New Zealand mining engineer named Charles Banks, who was considered to be the real expert on placer mining.

A prominent dredge design engineer named Frank Griffin who was already on the Board of Directors of Placer, contributed his personal engineering firm in San Francisco called Griffin Co., to provide engineering services to the project.

Attempts were made to get the New Guinea government to build either a railroad or road into the Bulolo area, but all failed which turned out to be a blessing in disguise. Instead, they turned their attention to obtaining air lift and ended-up using three Junker aircraft.

Most of the drilling after 1929, was done with a Ward, portable hand drill which due to its limited drilling depth, perpetuated the fiction of a shallow bedrock. The Bulolo flats were divided into a grid pattern by lines at right angles to the river 1,320 feet apart along each of which holes were spaced at 166 ft intervals.

Two Frenchmen, who had worked for Placer in Colombia on exploration, were sent in to do this drilling. The leader, L.A. Decoto, had 20 years of experience for such work and introduced new equipment and techniques of evaluation into New Guinea.

The Ward hand drill with a four-inch casing, was used because it was easily dismantled and made use of the one source of energy available, manpower. It slowed down, however, with depth and was difficult and tedious to penetrate beyond 40 feet. As the field developed, it gave the impression of a well-distributed placer over the whole of the river flats. Consequently, the distance between drill holes was increased and eventually, only every second line of holes was drilled.

From 1929, no real progress had been made on a railway or roadway. By then it was clear that the Bulolo area was "the world's most important placer deposit located since the Klondike." However, the area was wild and mountainous, heavily-saturated in disease and dysentery and the death rate was high among the miners and particularly the natives.

In the final report on the southern flats, it was estimated there were 39 million cuyd's of dredgeable gravel with a tenor of approximately 785 mg/cuyd. However, this turned out to be a gross under estimate particularly as to volume, because it was difficult to reach bedrock with the drills.

At the same time, however, the exploration manager had not been asked to find more extensive gravels, but only sufficient to make possible the floating of a public underwriting for the company. The other side of the coin was that had he known the full scope of the deposit, it might have appeared to be an "insurmountable problem" of raising capital; to design, build and transport large enough equipment to get into the area. Such an impression therefore could have resulted in the investors turning down the project.

Logistics Planning

Engineering studies were made and it was determined that it would be feasible to carry the dredges by air as long as no single component would exceed approximately 7,000 lbs and could fit into the fuselage of the Junker G-31 aircraft.

It was also at this time that the GRIFFIN CO. (predecessor to CPD), was assigned the task to carry out the engineering and design of the dredges and to be the technical support center for mobilizing the project.

A myriad of other arrangements had to be made, including the obtaining of water rights, installing a substantial plant for power generation, the building of an aerodrome, a large camp facility, and the entire logistics operation necessary to move in the dredges and the workers. Extensive feasibility studies had to be conducted in order to determine whether the project was realistic and economically viable.

The first two dredges were tailored to the conditions then known, which included a digging depth of 28 feet and an operating life of about 13 years. Each dredge weighed well over 1,000 tons. The bucket size was six cuft for the first dredge and 10 cuft for the rest of the dredges. The design of the dredges began in 1930 and the first dredge was in operation by March 1932.

Development Drilling

Further drilling was done in the general area nearby called Bulowat, using the Ward-type drill. Drilling was carefully done and holes went down as far as 50 feet or more. This expanded reserves and added a further 30 million cuyds on the Bulowat. The gold fineness appeared to be over 700, which was better than the Bulolo average. Further drilling expanded the yardage in Bulowat to an estimate of 60 million cuyds.

While the first two dredges were being designed in 1931, Banks was already discussing ideas of financing two additional dredges to dig the lower leases. In that same year, additional public funds were raised through the sale of stock to cover the cost of the additional dredges, which effectively expanded the capitalization of the company by 50%.

A provision was made that while selling 75% of the company's stock, all subsequent underwritings would maintain and not dilute, the 25% equity of Placer Development Ltd. Annual dividends were paid beginning in 1933, and continued at high levels as a bonanza to investors. The first underwriting, incidentally, was fully subscribed by one investor; BHP of Australia.

In the 1931 drilling of Bulowat, a sandstone bottom was found at a depth of 60 feet and it was seriously considered that the earlier Bulolo drilling had stopped at a false bedrock. By 1933, with the aircraft cargo lift, they were able to bring in powered drills of the Keystone-type and began re-drilling the lower Bulolo valley. Depths were found to exceed 120 feet in places with values continuing to those depths. The intermediate lenses that produced a false bedrock can be seen in the profiles of Figures 2 and 3.

Geological Theories

The results of the further drilling showed a topographical pattern in Bulolo similar to Bulowat, but at much greater depths. The valley was heavily overlain with gold-bearing alluvium, intersected by spurs which produced occasional areas of shallow bedrock. It was believed that the deposits must have been made by extreme action. If so, there would have been an extensive fault since the deposit was made, changing the elevations and gradient relative to the Bulolo riverbed. One of the theories advanced was that in early Tertiary times, much of New Guinea was submerged. Then massive earth movements occurred, accompanied by violent volcanic activity which affected the island and created the great mountain ranges. These movements may have continued until comparatively recent times.

It is thought that one source of gold occurred with mineralization in the early Tertiary period and this gave rise to metal of high quality with a fineness between 700 and 800. "Most of the gold, however, was probably derived from porphyries of the later Tertiary period of lower quality, fineness of about 500." This accounts for some of the erratic results in Bulolo. On the other hand, the single main source hypothesis (i.e. the gold deriving from an area of Upper-Edie only about one-square-mile in extent), explains the bulk uniformity of the gravels.

Dredging Systems

It was during this period of design development that the Pan-American jig was developed and patented by PLACER MANAGE-MENT LTD (renamed from GRIFFIN CO.), and installed on the dredges at Bulolo. Recovery of the gold was improved considerably when jigs were fitted on the dredges. Further gold saving techniques and systems were introduced and recovery (R/E) factor continued to improve. In addition, this improvement in recovery, allowed for profitable re-dredging of the tailings from the table/riffle dredges.

It is a testimony to the extensive planning, project management and technical interface with the engineering office of San Francisco, that the average operating time of the dredges exceeded 90%. Continuing experimentation of equipment and methods was conducted to constantly upgrade the performance of the dredges. Their production and recovery plus cross-fertilization between Bulolo and the Colombian property, further enhanced the results.

After some years of prospective exploration in surrounding regions of unknown country in New Guinea, which never resulted in mineable properties, the geological theory that the confinement of the gold was to that one river system from its origin was confirmed. Eventually, eight dredges were built and by November 1939, all were in operation in the Bulolo and Bulowat areas. Figure 4 Mining Leases on the Bulolo, and Figure 5 Dredge Courses, Bulolo and Bulowat, show the property and dredge planning.

In 1932-33, 4.5 million yd^3 that were dredged contained a tenor of 693.7 mg/yd^3 Au; but by 1940-41, when 19.5 million yd^3 were dredged, content had fallen to 414.7 mg/yd^3. It was noted that the advantages of having several dredges working simultaneously in the same general area had several beneficial results:

1. Reduced the overhead.

2. Resulted in standardization at a smaller unit cost of productions.

3. Extraction of the maximum quantities of gold while the price was rising and working costs were stable.

4. Access to the widest possible range of gravels; i.e. shallow, deep and marginal in value.

5. Ability to build up a substantial infrastructure/mining camp for maintenance, repair and logistics.

The centralized engineering office in San Francisco remained in an area of rapid communications and modern facilities to respond rapidly to all emergencies.

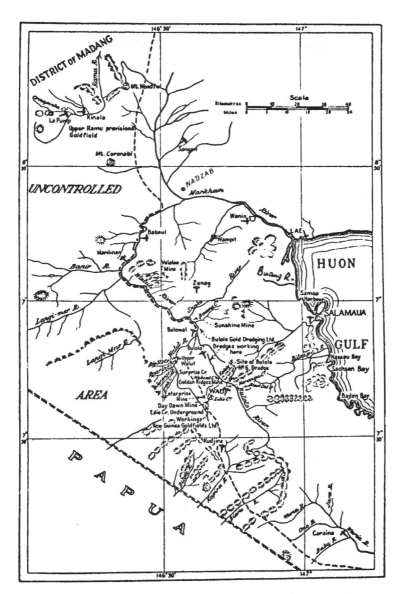

Map of goldfields, District of Morobe, 1934

Source: Fisher 1940

FIGURE 1-MAP OF GOLDFIELDS, DISTRICT OF MOROBE, BULOLO, NEW GUINEA, 1934

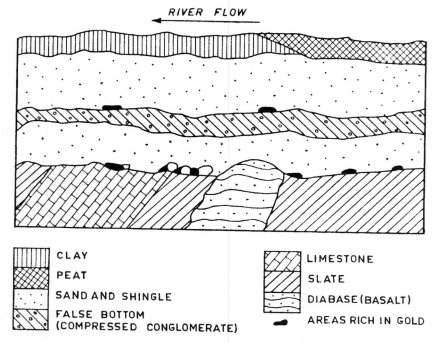

RIVER FLOW

CLAY

PEAT

SAND AND SHINGLE

FALSE BOTTOM
(COMPRESSED CONGLOMERATE)

LIMESTONE

SLATE

DIABASE (BASALT)

AREAS RICH IN GOLD

Ideal cross-section of a placer deposit (after Friedensburg)
showing a false bottom

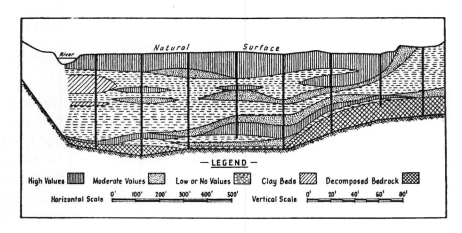

Natural   Surface

River

— LEGEND —

High Values   Moderate Values   Low or No Values   Clay Beds   Decomposed Bedrock

Horizontal Scale   0'   100'   200'   300'   400'   500'   Vertical Scale   0'   20'   40'   60'   80'

Typical cross-section of a Bulolo dredging area
**FIGURE-2 IDEAL CROSS-SECTION OF A PLACER DEPOSIT SHOWING FALSE BOTTOM.
FGURE 3-CROSS SECTION OF BULOLO RIVER.**

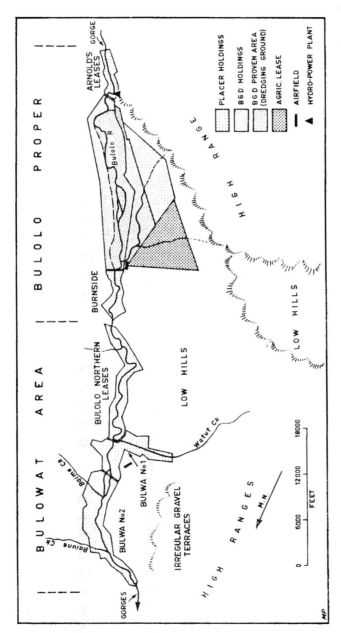

FIGURE 4-MINING LEASES ON THE BULOLO, JULY 1931.

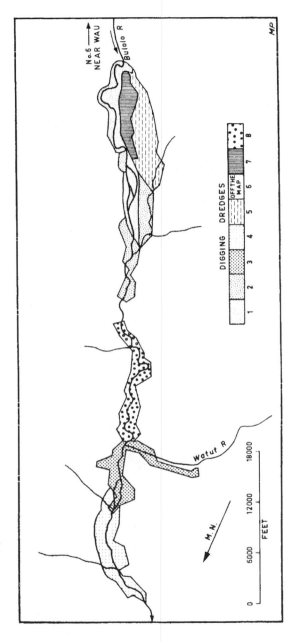

FIGURE 5-DREDGE COURSES, BULOLO AND BULOWAT

Case Studies

Appendix B
Geomorphological Study, Gold Placer,
Nechi River, Colombia

Background
In 1969, CPD on behalf of its parent PATO CONSOLIDATED PLACER DREDGING LTD., engaged Dr. Roy J. Shlemon to conduct a geomorphological examination of the company's gold dredging operations on the Nechi River, Antioquia, Colombia. This is a summary of Dr. Shlemon's report which demonstrates the benefits of such an examination, to an extensive gold placer deposit.

The Nechi River was first dredged in 1908 by a French company and later was acquired by CONSOLIDATED GOLD FIELDS of London. It was subsequently purchased by Placer Development Ltd. in 1932. In 1962, INTERNATIONAL MINING CORP. of New York, acquired PATO, which included CPD as its subsidiary and engineering center.

It was suspected that there were ancient paleo channels branching off the Nechi River, yet undiscovered which might result in an expansion of the existing reserves. Such turned out to be the case and, therefore, the study was considered to be highly successful in meeting its objectives.

Introduction
Economic geomorphology is a science still in its infancy. For many years mining geologists and engineers have employed basic principles of economic geomorphology by recognizing, for example, that the inside of meander loops probably contained more gold, platinum or other mineral deposits, than elsewhere in a stream.

Indeed, the widespread use of the terms "pay streak," "pay channel," and other colloquialisms clearly show that economic minerals, especially in alluvial deposits, are concentrated owing to changes in the hydraulic regimen of the stream.

Almost all major mineral-bearing streams were strongly affected by pluviality or tectonism during the Pleistocene Epoch; and it is the nature of this change that is studied by the geomorphologist. Thus the concentration of economically desired, heavy minerals in present and former stream channels, and the desire to recover these deposits creates a "wedding" of studies by the geologist, hydrologist and engineer; in essence, economic geomorphology.

In the long history of alluvial mining, little attention has been paid to the genesis of the sought-after deposits except where a "mother lode" or source area was located upstream. This of course is normal.

Few individuals or organizations have had the time, labor, or capital to invest in extensive theoretical studies about the origin of ores or alluvial deposits, except where there has been a direct and immediate effect on production. Most energy has been expended in more accurately delimitting reserves in presently producing areas or in increasing efficiency related to mining operations in general.

In recent years, several aggressive mining organizations have been profitably applying a combination of theoretical studies and technological advances to exploration. Some of the more common and better developed techniques include analysis of aerial photographs, in which patterns of vegetation or drainage offer a clue to the underlying structure; detailed study of infrared, thermal or other remote sensing imagery, the use of which, frequently supplements conventional photography; mapping of magnetic and gravitational anomalies, primarily using airborne equipment; and interpretation of subsurface structure, employing ground seismic equipment or analyzing well logs and cores.

All of these techniques, plus common sense, are tools of the explorationist whether he is a mining or petroleum geologist, archeologist, physical geographer, or geomorphologist. However, these tools and techniques, exotic and expensive, are often of little value unless the data derived are properly analyzed and interpreted. This is best accomplished when the field data are fitted into a theoretical framework concerning the genesis of the ores or deposits being sought.

The ultimate purpose of mineral exploration is to predict and to locate heretofore unknown deposits. However, this goal often is accomplished more effectively by analyzing presently known deposits. This is seldom done, for to most mine managers it appears economically irrational to spend time and capital analyzing existing deposits when the objective of exploration is to locate new ones. Yet continually reevaluating presently known reserves saves dollars in the long run.

This is especially true in alluvial exploration where it is often possible to predict deposits downstream only after the genesis of upstream minerals is known. Here, the tools and techniques of the explorationist provide the data for outlining the regional geomorphic history. In gold or platinum bearing streams for example, seismic or well log analysis provides data about the number, thickness and extent of presently exploited channels.

Coupled with knowledge about the theoretical environmental conditions that prevailed when the channel gravels were deposited, it is thus possible to predict more rationally the location of downstream mineral concentration.

The least benefit to be derived from this type of analytical study is a savings in drilling costs, for a geomorphological analysis frequently enables one to suggest more confidently the presence of a "pay stratum" in a particular location. Drilling may thus be concentrated in terrain more likely to contain the sought after mineral.

Geomorphological investigations are normally made in the early stages of exploration, followed by detailed engineering studies, and finally decisions of management. Nevertheless, geomorphologic study is not a "one-shot" affair, for new information is continually gathered as mining proceeds. This data supplements earlier geomorphic interpretations and ideally should"up-date" the study. Thus geomorphological analysis is fruitful at any stage of mine development.

This geomorphic study is one undertaken in 1969, well into the mature stage of alluvial mining on the Nechi River. The data derived probably would have been useful for planning exploration 10 or 20 years ago. Nevertheless, the principles formulated herein shed light on the genesis of the presently known gold-bearing gravels of the Nechi River, and the probable location of these gravels downstream. The Maps 1-6 Locational Map of Lower Nechi River Area, describe generally where this study is concentrated.

The following section briefly outlines the basic principles of economic geomorphology. Those principles particularly applicable to the lower Nechi River area are singled out for emphasis. The reason for this is two-fold:

1. Illustrate basic theoretical geomorphic concepts that underly the statements in this report concerning the evolution of gravel deposition and the concentration of alluvial gold.

2. Put into perspective those geomorphic principles that might prove to be of value to the readers of this report should further exploration be contemplated in tropical environments similar to the lower Nechi River.

Weathering and Transportation-
Geomorphic Applications to Economic Deposits

Broadly classified, most economic minerals are exploitable following weathering of the encompassing country rock, or after transportation and concentration by such agencies as running water, wind, waves or mass-wasting under the influence of gravity.

Residual deposits or those recovered at their point of origin, are of little concern in the lower Nechi River Valley and only brief mention is made of them. The transported or depositional minerals are by far the most economically significant deposits in the study area, and thus the geomorphic principles which govern their location are discussed in detail.

Residual Deposits

Most economic minerals were crystallized in a magma or were precipitated in hydrothermal solutions. Gold, platinum, copper, and many other naturally occurring elements are readily exploitedas native forms in lodes or combined minerals. These mineralsand the surrounding country rock are generally subject tomechanical and chemical weathering. This normally results in concentration of the minerals into residual deposits.

The Nechi River Valley alluvial gold undoubtedly came from at least several residual sources. A prime area is the Antioquian batholith, a complex, granitic-rock intrusive forming the "heart" of the northern Cordillera Central. The gold-bearing quartz veins that make up the lodes of the batholith have been exploited for many years.

Including downstream dredging, even today they yield about 60% of the gold recovered in Colombia. The Porce River, draining the northern part of the batholith, carried relatively coarse gold, most of it probably derived from the residual gold deposits. Many other streams also drain the batholith, generally carrying detrital gold eastward to the Magdalena River.

There is also very fine gold in the lower Nechi River. Some of this probably came directly from the Antioquian batholith, carried by the Porce and tributary streams. But a significant portion probably comes from hydrothermally emplaced, disseminated deposits in schistose rocks flanking the north and northwestern part of the hatholith.

The alluvial gold in the study area, therefore, most likely is a "mix," derived from many residual deposits in the adjacent highlands, emplaced by at least two different methods.

Transported Deposits

Economic minerals transported to their final place of concentration are generally termed "alluvial deposits." Most of these deposits are "heavy minerals;" i.e., minerals with specificgravity generally greater than 3.0 and thus naturally segregating when stirred or mixed by a transportational agency such as wind, waves or running water.

Historically, the more famous and significant alluvial or placer deposits have been gold, platinum, and tin-bearing rocks. Diamonds and other gem stones also have been placer mined. Rutile, ilmenite, and zircon are other relatively heavy minerals naturally concentrated into workable deposits, especially on coasts in the form of beach placers.

The most common natural transportational agencies are wind (eolian processes), glaciers, shoreline currents, mass-wasting (gravitational processes), and stream (fluvial processes). Sediment movement by streams, past and present (fluvial processes), is by far the most important gradational process affecting evolution of the geomorphic landscape and the one most likely to concentrate alluvial economic minerals.

This is especially true for high specific gravity minerals such as gold and platinum. Fluvial processes clearly have strongly affected the concentration of gold along the lower Nechi River. Hence, in the following sections, the geomorphic significance of fluvial processes is singled out for detailed analysis.   For perspective, the other gradational processes are mentioned briefly, though they are not of great significance in the Nechi Valley.

Coastal Processes/Beach Placers

The same streams that presently carry gold, platinum, tin or other heavy minerals, eventually debouch directly or through tributaries into the sea. Here the natural "riffle" action of longshore currents may concentrate these heavy minerals into economic deposits, often exploited when demand and extractive technology permit.

Frequently associated with "black sands" (high specific gravity, iron-magnesium bearing minerals), these deposits are worked by methods ranging from primitive hand washing to gravity and magnetic separation, in bucket-line or suction dredges.

The nature and location of beach placers is beyond the scope of this study; however, gold and platinum bearing rivers of western Colombia, especially the swift waters within the Department of Choco, have been carrying heavy minerals to the Pacific or Caribbean since at least Pleistocene time.

There, concentrated by longshore currents, economic deposits undoubtedly exist; some have been worked, others, most likely, will be exploited when more precisely located and analyzed following detailed geomorphological survey and application of direct and remote sensing techniques.

Eolian (Wind) Processes

The wind, as other geomorphic agents, erodes and deposits. In present arid and semi-arid regions, winds are known to shift fine-grained sediments in a preferred direction, occasionally leaving heavier, economically desireable minerals behind.

Concentration of economic minerals by wind is relatively small and is of little concern in the lower Nechi River area. Certainly the present climatic regime is not conducive to significant eolian action and it is not likely that winds during drier, Pleistocene interglacial climates gave rise to concentrations of alluvial deposits.

Glaciation

Though not directly affecting the lower Nechi River area, glaciation in the Andes during Pleistocene time undouhtedly caused many streams to carry large, glacially-plucked cobbles and boulders much farther downstream than they do today. The increased competence of the glacial streams permitted them also to carry and concentrate gold and other high-specific gravity minerals that previously had been exposed and weathered in the adjacent highlands. It is thus very likely that the Nechi River basal channel gravel, for example, the most sought auriferous unit, was deposited when the Central and Western Cordillera of Colombia were glaciated in Pleistocene time.

Mass-wasting (Gravitational Processes)

Mass-wasting generally refers to the downslope movement of rocks and sediments under the influence of gravity. Some movement may be rapid, such as landslides and rockfalls, or very slow and imperceptible, such as soil creep. The rapidity of movement, to a great degree depends on the texture or particle size and the water content within the sediments.

Mass-wasting generally only indirectly or locally affects concentration of economic minerals. This is exemplified in the lower Nechi River Valley where some large granitic boulders, four feet in diameter, have been dredged approximately 60 feet below the present surface, about 20 miles downstream from the nearest canyon outcrop of granite rock.

Only a stream of very high competence could have transported these boulders; even thehighest modern floods do not move boulders of this size.Geomorphological studies of the Medellin and adjacent valleys, 80 to 100 kilometers upstream from the large boulders, show that, in the past, huge landslides have repeatedly dammed the main channel and many tributaries of the upper Nechi and Porce Rivers (Shlemon, 1970).

It is likely that these natural dams were eventually topped or burst days or weeks later releasing "catastrophic" waves of water downstream which, in turn, carried mass-moved boulders far out into the flood-plain of the lower Nechi River. These landslide controlled streams transported gold and other high specific gravity minerals farther downstream than would have occurred during "normal" channel flow.

Local mass-moved, concentration of minerals can also occur when minor slumping of terrace gravels along a river bank "redeposits" gold. Immediately downstream from the minor landslide or slump is often a likely place for an abnormally high gold concentration.

Running Water (Fluvial Processes)
The best known and most important geomorphic process is running water. Stream placers have yielded the greatest quantity of gold, heavy minerals, and precious stones from time of the "ancients" to the present. Running water is an excellent medium for separating light and heavy minerals.

With ever-changing velocity, a stream transports and deposits heavy minerals irregularly and in widely separated places along its course. Subject to sudden changes in velocity owing to floods, a stream can carry gold many miles in a short time, then flow with reduced velocity for miles downstream transporting only fine-grained sediments such as silt and clay.

The most favorable places for accumulation of gold is where local stream velocity is suddenly reduced. Represented graphically in Figure 1, this occurs commonly along the inside of shifting meander loops, or where a stream flows over irregular bedrock. Gold concentrated naturally by these processes has heen known for years, and has been exploited by miners and prospectors to recover local "bonanzas" in California, Alaska, and in other well-documented auriferous areas.

Local accumulation of gold may occur at any place; over the long run, however, the greatest concentration is generally found in gravels just below canyons or at points where the stream gradient is greatly reduced. Fine or flour gold may he carried far downstream, but the largest nuggets and highest values-per-unit volume invariably are found in the low "foothill" regions.

This is as true along the lower Nechi River today as it was a hundred years ago along the great alluvial placers of the western slopes of the Sierra Nevada mountains in California (Hutchins, 1905, Page 1010).

These elementary principles of heavy mineral concentration, applicable to modern gold-bearing streams, are likewise true for "old streams;" that is, gravels, sands and silts laid down by ancestral channels of the present drainage. The same fluvial processes that influence modern deposition were also operable during Pleistocene time, although on a different scale.

This point is especially pertinent to exploitation of the lower Nechi River, for much of the gold presently recovered is "old," laid down thousands of years ago. An important task of the economic geomorphologist, then, is to reconstruct fluvial conditions ofthe past. The principles governing ancestral stream deposition are briefly treated in the following section.

Regional Geomorphic Evolution

Influence of Fluvial Processes

Emerging from steep canyons, passing through the foothills, and onto relatively flat valley plains, streams almost invariably deposit coarse-grained sediments first (most of these deposits are gravels). With the gravels, are laid down the smaller but high specific gravity minerals such as gold. Thus is developed the association of channel gravels and gold.

Tracing the auriferous gravels downstream is relatively simple, although frequently expensive. Holes are generally drilled in a grid pattern adjacent to the known gravels, and gold values are computed from core samples recovered. In this way a "pay channel" is often discovered and followed by the dredge.

From a geomorphic point of view, the so-called pay channels are really gold deposits laid down at one particular time in the history of the stream. Why and where this gold is deposited generally depends on the hydraulic competence of the stream, its gradient, and the then existing local "restrictions," which reduced water velocity and allowed the gold to settle out.

Reconstructing the paleo-fluvial history of the stream is the task of the geomorphologist. This analysis offers two immediate benefits to alluvial mining: Firstly, it enables one to predict likely locations of pay channels downstream, Secondly, it can reduce exploration costs because close grid pattern drilling is not necessary in those areas less likely to have significant auriferous gravels.

There is abundant geomorphic evidence suggesting that there were "sudden" changes in the hydraulic competence of the lower Nechi River. The main line of evidence is the presence of numerous flat surfaces or terraces which flank both sides of the Nechi River and many of its tributaries.

A terrace is simply an abandoned floodplain, or the old level of the stream. A fundamental question is why the Nechi River suddenly cut deep new channels leaving its former course as terraces? The question, to some extent, is answerable by analyzing the types of terraces found on the Nechi.

There are two basic types of terraces, described in Figure 2. Both occur along the lower Nechi River. One is the "cut" or strath terrace. It usually forms when a stream is stable or in equilibrium, neither significantly downcutting or depositing. At this time the channel may meander across its floodplain cutting into flanking bedrock or older alluvial deposits. When the river is rapidly incised, the cut surfaces are left above the new channel as terraces.

A second terrace is the so-called "fill" type. It forms when a stream deposits sand, silt, gravel, and other heterogeneous sediments along its main course, floodplain, or natural levee, followed by relatively rapid downcutting. The old alluvial deposits now underly the abandoned floodplain, but perhaps may be left many feet above the new drainage. Frequently, the old fill deposits are auriferous. This is exemplified on the lower Nechi River where hydraulicking continues to exploit old channels now high above the present stream.

Two interacting influences seem to explain best the origin of the Nechi River terraces: (1) The direct effect of tectonism or uplift, which changes stream gradient; and (2) The indirect effect of climatic change, altering the hydrologic regimen of the stream, and world-wide sea levels. The relationship of these two influences on Nechi River terraces is briefly reviewed in the following section.

Tectonism and Terrace Formation
Uplift in the headwaters or subsidence along the lower course will cause a stream to suddenly deepen its existing channel. The old course is then left as a fill or cut terrace. The terrace may be paired, that is, found at similar elevation on both sides of the new channel, or it may

be a singular feature, exposed discontinuously downstream along one side or the other of the present stream. New and former gradients of streams subject to tectonism are illustrated in Figure 3. Invariably the older channels, perhaps preserved as a terrace, has a higher gradient than the present stream.

Whether abandonment of the floodplain and creation of the terrace results from uplift or from subsidence, the effect is the same; former channel gravels, perhaps auriferous, are vertically displaced with respect to the present or modern stream sediments. At only one place, the "hinge" point (Figure 3), will the fluvial deposits of two different ages coincide. This is one place where "above average" gold recovery may be expected.

Climatic Change and Terrace Formation
The indirect effect of climatic change on terrace formation is much more complicated and uncertain than the more obvious direct influence of tectonism. Nevertheless, during regional climatic change in Pleistocene time, many streams throughout the world cut new channels, abandoning old courses and leaving the contained deposits as fill terraces.

Glacial meltwater in many high latitude or high altitude streams profoundly affected the hydrolgic regime of rivers, especially the delivery of water and sediment. Many streams whose headwaters were not directly glaciated nevertheless similarly underwent channel change owing to increased rainfall (pluviality). Also, glacial or pluvial climates directly affected the density and type of natural vegetation; this in turn influences sediment yield to the rivers. For those streams that debouch directly into the ocean, Pleistocene climatic change had still another effect. During glacial time worldwide sea levels (ultimate base level), were lowered, possibly 400 feet or more. Streams then cut to the new level, abandoning their old channels and creating terraces. During this interglacial time, the sea rose above its present level, 100 feet higher than sea level today.

Stream gradients were then reduced; hydrologic competence was similarly decreased. By climatic change alone, terraces may thus be formed. epending on the quantity of available water and sediment load, a major stream, such as the Nechi system, throughout Pleistocene time, alternatively deposited sand and other fine-grained sediments and gravels. Thus auriferous gravels of the Nechi River were deposited in "pulses;" that is, most of the gold was not laid down continuously, but rather was transported and deposited during specific, limited periods.

Gold-gravel accumulation on the lower Nechi River is probably related to Pleistocene climatic change, with most deposition occuring during each glacial (pluvial) time. During interglaciation, worldwide base level (sea level) rose, the streams thus had lower competence owing to reduced water availability and lower gradients. As a result, fine-grained sand and silt, rather than gravels and gold, were deposited.

Although certainly not active as long as natural processes, man as a geomorphic agent, has dramatically and rapidly changed the land surface. In mining, this is very obvious where hydraulicking continues to wash millions of cubic yards from high level benches or terraces. Man is also affecting the landscape in another, albeit less obvious way. He is causing downstream sedimentation by upstream agriculture and mining.

Agriculture and related activities, such as plowing and burning of the natural forest, certainly increase runoff and sediment yield. This has been demonstrated throughout the world. Similarly, hydraulicking and sluicing of old channels and terrace gravels washes tons of sediments into present streams. Most of the new sediments are fine-grained and thus carried far downstream. Eventually deposited, these sands and silts build up the channel of a stream and eventually spill onto the floodplain. The base of the stream becomes higher; in essence the frequency of floods increases.

Accelerated sedimentation due to mining was first described by Gilbert (1917) for the Sierra Nevada, California. This classic study was cited by many anti-mining interests in support of legislation restricting hydraulicking. Yet today, sedimentation due to mining is continuing unabated on the lower Nechi River. But the deleterious effect of hydraulic mining in this rather remote section of Colombia is almost unknown and of little concern to the general public.

Hyraulicking has other undesirable aspects, some which affect other forms of mining. For example, the above-average, man-induced sedimentation in some 400 odd years of gold exploitation has caused very rapid burial of old Nechi Riverchannels (low terraces) many milesdownstream from the hydraulicked area. This is especially true near the "hinge point" or places where gradients of the modern stream and terraces converge (Figure 3). Here, former low terraces or henches are now buried by more than 20 feet of "man-deposited" alluvium, most of which probably came from upstream hydraulicking. Hence, instead of recovering gold from old terraces just a few feet above or below the present river level, dredges must now excavate many feet of barren overburden sand and silt.

Hydraulickly-mined induced sedimentation on the lower Nechi River is not a new phenomenan. As early as 1913, William Ward, describing operation of the old Pochet steam powered dredge, noted that former red (oxidized) bench gravels were often encountered 15 feet below the surface; this due primarily to sedimentation since the time of Spanish mining operation. Holes drilled 40 years ago in the Jobo Lake areapenetrated oxidized terrace gravels now 20 feet below the surface. Those terraces or low benches, formerly above the Nechi River, are being covered by an increasing thickness of sediments downstream with the passage of time.

See Table A Auriferous Gravels of the Lower Nechi River Valley Near El Bagre, for an overview of geological epochs and their influence.

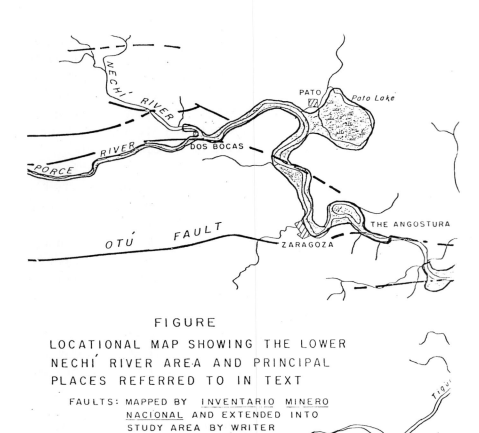

FIGURE

LOCATIONAL MAP SHOWING THE LOWER
NECHÍ RIVER AREA AND PRINCIPAL
PLACES REFERRED TO IN TEXT

FAULTS: MAPPED BY INVENTARIO MINERO
NACIONAL AND EXTENDED INTO
STUDY AREA BY WRITER

BASE: EL BAGRE AND LIBERIA QUADRANGLES

N

SCALE: 1:100,000

0 1 2 3 4 5
KM

AREAS FOR
DREDG.
1970

FIGURE 1-LOCATIONAL MAP OF NECHI RIVER AREA

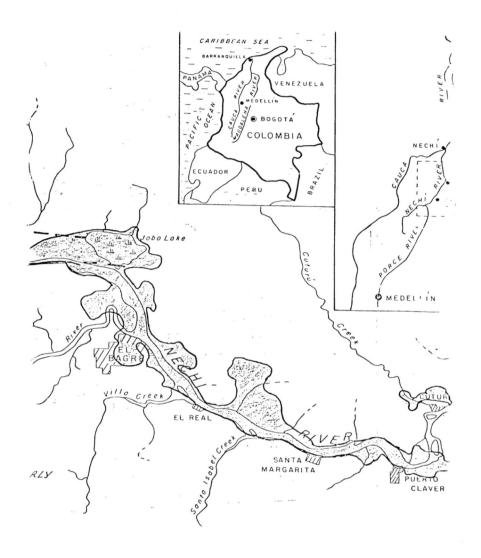

FIGURE 2 -NECHI RIVER AND REFERENCE MAPS OF COLOMBIA AND MEDELLIN

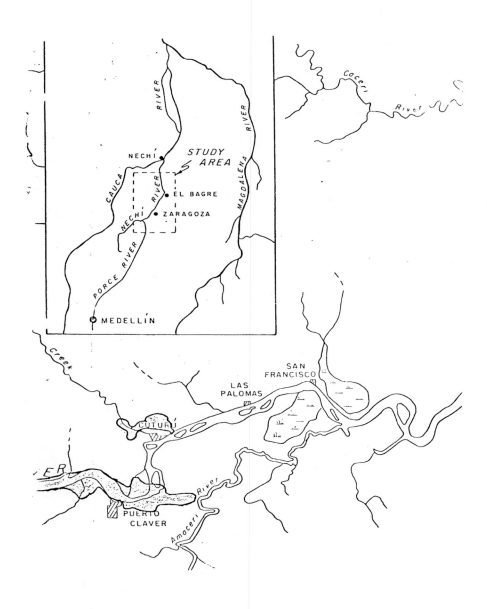

FIGURE 3 -LOCATION MAP OF STUDY AREA AND TRIBUTARIES TO NECHI RIVER

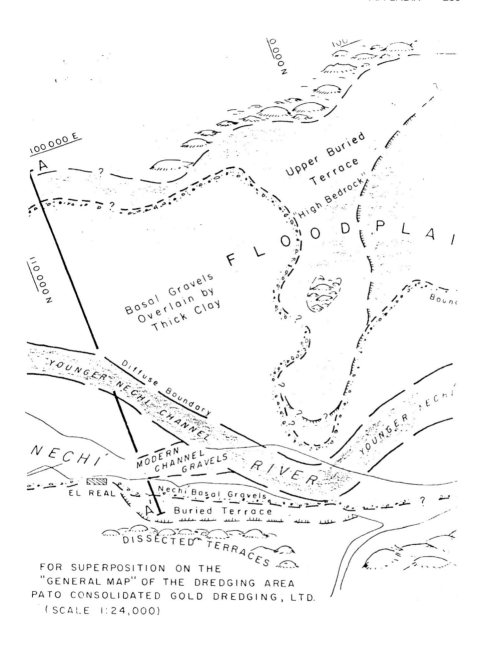

FOR SUPERPOSITION ON THE
"GENERAL MAP" OF THE DREDGING AREA
PATO CONSOLIDATED GOLD DREDGING, LTD.
(SCALE 1:24,000)

FIGURE 4 - NECHI RIVER CHANNELS AND FLOODPLAIN

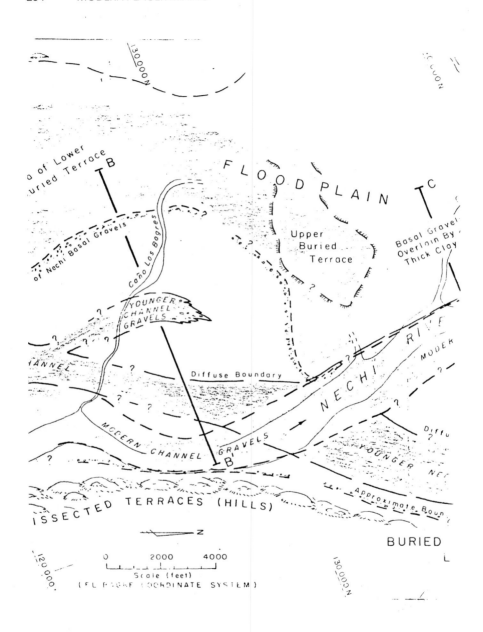

FIGURE 5-EXTENSION OF FIGURE 4, NECHI RIVER TERRACES

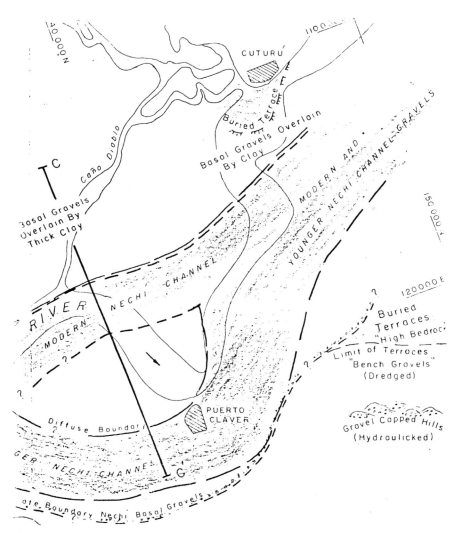

CUTURÚ

Buried Terrace

Basal Gravels Overlain
By Clay

MODERN AND
YOUNGER NECHI CHANNEL GRAVELS

Basal Gravels
Overlain By
Thick Clay

Caño Diablo

40.000 N

1100

150 000

120000 E

R I V E R

MODERN

NECHI   CHANNEL

Buried
Terraces
"High Bedroc'
Limit of Terraces
"Bench Gravels"
(Dredged)

PUERTO
CLAVER

Diffuse Boundary

Gravel Copped Hills
(Hydraulicked)

GER   NECHI CHANNEL   G

ate Boundary Nechi Basal Gravels

BURIED TERRACES AND CHANNEL GRAVELS OF THE
LOWER NECHI' RIVER FLOODPLAIN

FIGURE 6 -EXT FIGURES -4 & 5 BURIED TERRACES AND
CHANNEL GRAVELS OF LOWER NECHI FLOODPLAIN

(A)

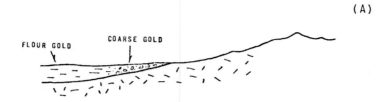

FLOUR GOLD   COARSE GOLD

(B)

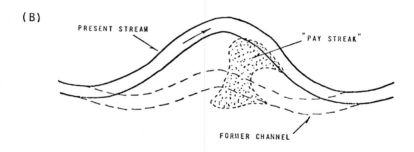

PRESENT STREAM   "PAY STREAK"

FORMER CHANNEL

Fig. 1 - (A) Generalized diagram showing gravel and coarse alluvial gold
deposition near the mountain front and fine (flour) gold
carried out into a valley flat.

(B) Typical gold concentration along the inside of a migrating
meander loop.

**FIGURE 7 -(A) DIAGRAM OF GRAVEL & COARSE ALLUVIAL GOLD DEPOSITION
(B) TYPICAL GOLD CONCENTRATION ALONG INSIDE OF MIGRATING MEANDER
LOOP**

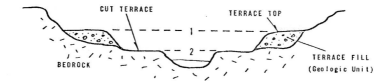

Fig. 8 - Hypothetical paired cut (strath) and fill terraces. The fill terrace (number 1) is the oldest. The stream cut a surface and later deposited gravels and fine-grained sediments. Subsequent incision led to cutting of a younger terrace (number 2). The term terrace refers to geomorphic surfaces; underlying rocks are geologic units.

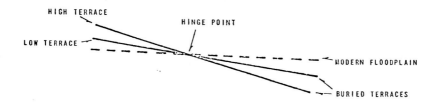

Fig. 3 - Downstream convergence of terrace and modern stream gradients.

FIGURE 8-HYPOTHETICAL PAIRED CUT AND FILL TERRACES
FIGURE 9 DOWNSTREAM CONVERGENCE OF TERRACE AND MODERN STREAM GRADIENTS

TABLE A

Auriferous Gravels of the Lower Nechi River Valley Near El Bagre

Inferred Age	Gravel Unit	Notes
Holocene; 0–2,000(?0) years ago	Modern Channel Gravel	Generally underlies present river; slightly auriferous
Holocene; less than 10,000 years old (neoglaciation?)	Nechi Younger Channel	Confined to present floodplain, interbedded gravel and sand; slightly auriferous
Late Pleistocene; Wisconsin glacially derived deposits	Nechi Basal Channel	Most extensive channel; high aureferous, especially near bas undulating contact with Tertiary bedrock (pena)
Late Pleistocene; pre–Wisconsin	Low Terrace (Deep Buried Terrace)	Generally less than 30 feet abov present floodplain; locally dredged; buried downstream
Late Pleistocene; pre–Wisconsin	High Terrace (El Bagre Surface; Upper Buried Terrace	Hydraulically mined; buried down-stream; restricted course of younger channels
Late Tertiary (?) or early Pleistocene	Nechi Pediment Gravel	Hydraulically mined near Zaragoz planed across schists and Tertiary sediments; faulted
Tertiary	Auriferous Channels and Silt Deposits	Discontinuous gravel-filled channels; quartzitic cobbles abundant; little gold under Nechi floodplain; hydraulically mined upstream near Porce Cany

TABLE A
AURIFEROUS GRAVELS OF THE LOWER NECHI RIVER VALLEY NEAR EL BAGRE

CASE STUDIES

Appendix C
Exploration of Au and Pt, Choco, Colombia

Background
A target area for gold and platinum exploration was selected in Choco, Colombia. The area selected was a flood plain covering 3,700 acres. Access to the area was by 14-foot fiberglass boats powered by 40hp outboards, running two hours of travel from the nearest camp.

Ward Drill
Because of the access problems and high availability of inexpensive labor, two Ward hand drills were used for exploration. The exploration crew consisted of 15 men for each drill; one engineer, one surveyor and two chainmen. All personnel were hired locally.

The initial drilling program consisted of 485 holes averaging 38.5 ft in depth. The holes were placed by surveying a grid on a base line selected from aerial photographs. The drill lines were 1,600 feet apart, with a hole spacing of 200 feet along each line. As a result of the initial drilling, a reserve of 47 million yd³ with an average of 0.464gr (grains) Au/yd³ and 0.471gr Pt/yd³ was established. Figure 1-Drilling Program Map-Choco, shows dotted lines of drill sites and grid spacing.

This portion of the drilling required two years for completion. The decision was made to continue the drilling program on a closer spaced grid, in an effort to delineate a more continuous channel of values. The second set of drill holes was set up on the same base line as the first, and the drill lines were surveyed in half way between the first set of lines (800 feet).

reduced to 40,525,000 yd^3 and the grade had increased to 0.647 grAu/yd^3 and 0.581 grPt/yd^3. This portion of the drilling consisted of 273 holes averaging 39.2 ft and took a little less than two years to complete. An area of 702 acres contained the reserves of the second drilling program. When averaged over the four year period, production from the Ward Drill was 280 ft/month at a cost of $3.00/ft.

Evaluation of Boreholes by 4" Ward Hand Drills
The basic idea in evaluating a borehole sample is to estimate as reasonably as possible the average content of mineral-per-unit volume of sample from drilling.

The Ward drill utilized a 4-inch inside diameter casing in the Choco for transportability into the difficult, swampy and heavily forested areas along the main rivers and creeks. The method of calculating a borehole sample was hased on factors derived from the size of the drive shoe, the inside diameter of the casing and the weight of metal recovered from the hole. The drive shoe factor is the theoretical length of drive required to cut a sample volume containing one cubic yard.

It is at best, theoretical because in practice neither is the drive shoe a keen edged instrument cutting out a core of uniform character, nor is the material sampled a homogeneous mass capable of being bored in this manner. It therefore follows that the best that can he attained is the use of a factor that will approximate accurate sampling, this accuracy being checked only after the borehole is finished.

In the Choco, the theoretical factor, using the total area of a new shoe, worked out to be 188 feet. In the field a factor of 200 feet was used. This, in part, compensated for the wear of the shoe's edge. The core data within the limits of accuracy with which it is obtained, a correction factor of adjusting the drive shoe factor from a variable to a constant is possible.

To illustrate, consider both the 200-foot factor and the 243-foot factor for the same shoe when worn out (disregarding the question of swell for the moment); see Figure 2-Straight Thread-Type Drive Shoe, for new and worn. The interior area of the casing is 12.57 in^2 (4-inch diameter). The area of the drive shoe corresponding to the 200-foot factor is 19.44 in^2. To determine the theoretical core rise in the pipe for the corresponding drive:

Drive : Rise :: 12.57 : 19.44

Therefore, for a one-foot drive, the theoretical rise is 1.54 ft.
For the 243-ft. factor, the corresponding drive shoe area of 16 in^2 and the area of the interior of the casing remaining the same, the following derivation results:

Drive : Rise :: 12.57 : 16

Therefore, for a one foot drive the theoretical rise is 1.27 feet.

From which it is clear that the introduction of a core factor correction compensates the differences due to the different shoe sizes:

200:243::1.27:1.54

Even if the core measurements are not so accurate in the field,these are more precise than the straight application of a theoretical factor, which is variable.

The earlier drill logs in the Choco show that the amount of core varies between wide limits depending on the ground character, water, and the experience of the driller. The material which enters the tube and is measured therein and after it has been pumped out, must be taken into account for the correct calculation of a borehole sample. The use of the core data does eliminate variations due to erratic cores and, in consideration of a given area, or set of areas, tends to reduce the drill hole data to a common base for the various holes involved.

In most cases of land drilling in the Choco, the gravel is not so coarse and the cores are below normal. That is, the rise of the core measurement in the calculation tends to increase the expectant grade. In river drilling of coarse gravels where there is a greater water pressure on the outside of the casing, the cores are noteably greater. The adjustment, in this case, tends to lower the indicated grade.

The following drill logs are typical showing the method of adjustment taking the core data into consideration. The factor of swell of core as it enters the pipe is thus introduced. There is no definite basis on which to determine the amount of swell, but this is not too important a factor to consider for swell is not very great inside the casing, since: (1) the driving down of the shoe tends to compact the material at the point of the shoe and as it enters the shoe; and (2) the material is forced into a pipe of a smaller cross-section than that of the cutting shoe area. For the purpose of the calculation of the logs, a 5.0% swell factor was applied as an average for the column.

The desired rise to obtain an ideal core from a drill shoe using the 200-ft. factor is then:

1-ft. drive:Rise::1:(1.54 + (5% x 1.54))

The rise then is 1.62 ft. for a 1-ft. drive.

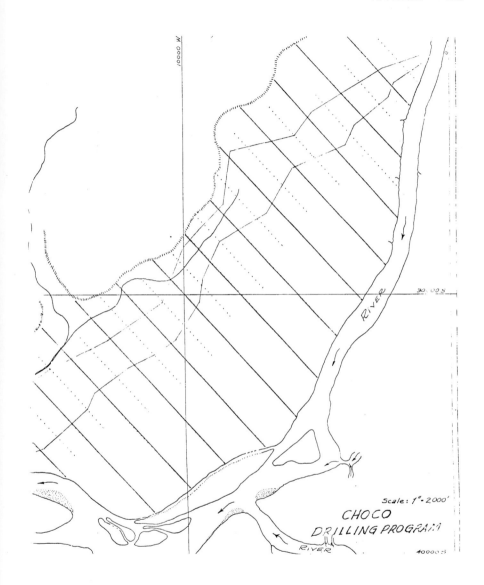

FIGURE 1-CHOCO DRILLING PROGRAM MAP

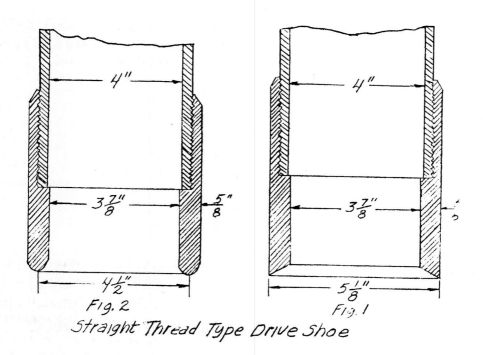

Fig. 2

Fig. 1

Straight Thread Type Drive Shoe

FIGURE 2-STRAIGHT THREAD TYPE DRIVE SHOE

Original ~~Duplicate~~ Eng. From. 20

Pato Consolidated Gold Dredging, Ltd. FIELD LOG

Examination [illegible] Line 164.1 Hole [illegible] Sheet 1 of 2 Sheets

Elevation Collar 69.? Co-ord. N 115.439, E [illegible] Date Started [illegible] 19[illegible]

Offset from Stake, Bear ____ ; [illegible] ft hor., [illegible] ft. vert. Date Compl. [illegible] 19[illegible]

Pump. Time Hr. Min.	Depth Drilled ~~Ft.~~ m	Depth Drilled 1/10	Core: Drive	Core: BEFORE Pumping	Core: AFTER Pumping	Colors 1	Colors 2	Colors 3	Colors 4	Meas Vol ~~Cu. Yds.~~	Corrections Mgs.	Est. Wt. Mgs.	Formation
[illegible]	[illegible]		[illegible]										
[illegible]	4	[illegible]	4 [illegible]	[illegible]						[illegible]			[illegible]
[illegible]	4	[illegible]	1 20	[illegible]						[illegible]			
[illegible]		[illegible]	1 30	1 [illegible]						[illegible]			[illegible]
1:3?	[illegible]	10	1 [illegible]	[illegible]				1		.004	+0.1	0.3	[illegible]
2:10	9	[illegible]	1 20	41						.006			[illegible]
[illegible]	10	5	90	5?						[illegible]			
[illegible]	11	20	[illegible]	[illegible]						.015			
3:45	12	[illegible]	[illegible]	1 31						.01?			
1:00	13	[illegible]	1 2?	1 5?						[illegible]			
[illegible]	15	[illegible]	20	1 [illegible]						[illegible]			[illegible]
1:2?	15	[illegible]	90	1 11						[illegible]			Clay [illegible]
2:30	16	[illegible]	[illegible]	51						[illegible]			
[illegible]	17	10	60	51						.007	+0.1	0.3	
[illegible]	17	70	[illegible]	[illegible]	1			2	2	.11	/	0.7	[illegible]
10:?2	[illegible]	[illegible]	[illegible]	[illegible]	56			1	2	[illegible]	/	0.3	[illegible]
11:??	[illegible]			1 ?0	4?			2	8	[illegible]	+0.1	0.9	
12:20	19	[illegible]	60	[illegible]	78					.012			
[illegible]	19	[illegible]	30	[illegible]	1				15	[illegible]	/	0.2	
[illegible] 1:35	20	[illegible]	[illegible]	2 64	31				14	.010	/	0.2	
2:?5	20	40	30	1 [illegible]	39				12	.008	-0.1	[illegible]	
[illegible]	21	71	[illegible]	[illegible]	21			20	[illegible]	[illegible]	-1.9	8.3	[illegible]
[illegible]	21	[illegible]	[illegible]	[illegible]	[illegible]			75	[illegible]	.007	-8.2	[illegible]	
11:20	21	[illegible]	[illegible]	1 [illegible]	[illegible]			50	370	[illegible]	-2.4	[illegible]	
[illegible]	21	[illegible]	[illegible]	1 11	6?			62	648	[illegible]	-7.7	[illegible]	
12:30	21	[illegible]	[illegible]	[illegible]	[illegible]			35	[illegible]	[illegible]	+1.8	22.?	
[illegible]	22	21	30	75	[illegible]			18	330	[illegible]	-0.4	[illegible]	
[illegible] 9:?5	22	[illegible]	[illegible]	71	52			[illegible]	[illegible]	[illegible]	+1.7	21.2	
[illegible]	23	[illegible]	30	57	9			[illegible]	[illegible]	.00?	-1.1	11.?	
11:40	[illegible]	1?	[illegible]	[illegible]	12			[illegible]	58	.007	-1.0	[illegible]	
2:00	23	40	30	63	18			[illegible]	[illegible]	.005	+0.1	[illegible]	
3:30	23	70	30	78	21			7	28	.002	-1.2	[illegible]	
[illegible]	24	[illegible]		33	12			4	15	[illegible]	/	[illegible]	
[illegible]	24	[illegible]	3	57				5	[illegible]	[illegible]	-0.2	2.1	
11:[illegible]	24	[illegible]	3	[illegible]				[illegible]	[illegible]	[illegible]	-0.1	[illegible]	

DRILL *
- Type & No. [illegible]
- Size Drive Pipe [illegible]
- Dia Drive Shoe [illegible]

TIME LOG *
- Moving
- Drilling
- Pulling
- Delays
- Total

DEPTH, ETC. *
- Water Level
- Overburden
- Gravel
- To Bedrock
- Penetrated Bedrock
- Total Drilled
- Type Bedrock

CALCULATIONS
- Calc. Vol.
- Meas. Vol.
- Core Vol.
- Drive Shoe Factor
- Core Factor
- Vol. Factor
- Est. Wt. Mgs. *
- Wt. Gold. Mgs.
- Correction
- Corrected Gold, Mgs.
- Est. Fineness
- U. S. $ per Fine Oz
- Wt. Black Sand Oz
- Bedrock elev.
- Normal W. L.
- Dredging Depth
- Below Normal W. L.

SEE PAGE #2

PERSONNEL
- Driller *
- Foreman * [illegible]
- Calc. by
- Checked by
- Engineer in Charge

* These entries must be completed in the Field.
See over for Remarks
B. R. Class No.

Estimated Mean Value: U. S. cents per Cubic ~~Yard~~

* 240 [illegible]

Pay Stratum ____ ft. to ____ ft = ____ cents.
Tailings ____ ft. Virgin Ground ____ ft. Bedrock ____ ft.
Calculated Total Dredging Depth (excl. water) 24.90 m ~~ft.~~ ____ cents, (____ mgs. per cu [illegible]
LINE ____ HOLE ____

FORMA 94 - ENERO 73 500 Tip. Ideal

FIGURE 3-DRILL LOG 1

Original ~~Duplicate~~

Eng. From 20

Pat~ Consolidated Gold Dredging, Ltd.   FIELD LOG

Examination _Nedli + 13_   Line _1844_   Hole _C97_   Sheet _2_ of _2_ Sheets

Elevation Collar _60_   Co-ord. N _8,400_ , E _10,800_   Date Started _22_ _Mayo_ _22_ 19 _74_

Offset from Stake, Bear _____ ;____ ____ft hor., _3.00_ ft. vert. Date Compl. _11:00_ _Junio_ _7_ 19 _74_

Pump. Time		Depth Drilled		Core			Colors				Meas Vol	Correc- tions	Est. Wt.	Formation
Hr.	Min.	Ft.	1710	Drive	BEFORE Pumping	AFTER Pumping	1	2	3	4	Cu. Yds. m³.	Mgs.	Mgs.	
1:05	24	20		25		-50					.1			Yellow ... clay ..
1:10	25	20		30		-35					.005		1.23	
				Klean										
				Clot Coloid				2:45						
1:5				Finished Drilling										
1:50 all				Finished Pulling										

Estimated Mean Value: U. S. cents per ~~Cubic Yard~~ _____

Pay Stratum_____ft. to._____ft._____cents.

Tailings_____ft. Virgin Ground_____ft. Bedrock_____ft.

Calculated Total Dredging Depth (excl. water) _24.95m_ ~~ft.~~ __ _7.4_ cents, (__ _.747_ mgs. per cu ~~yd.~~ m³.)

Plus corrections not used

LINE _1844_   HOLE _C281_

Ward No. 46

DRILL *

Type & No. _KJDD #10_

Size Drive Pipe ____ _42_

Dia Drive Shoe ____ _5 3/8_

TIME LOG *

Moving _____

Drilling _2_

Pulling _____

Delays _12:40_

Total _2:50_

DEPTH, ETC. *

Water Level+ _5.7 lls. graves_ 7.4

Overburden _14.50_

Gravel _2.95_

To Bedrock _24.15_

Penetrated Bedrock _____

Total Drilled _____

Type Bedrock _____

CALCULATIONS

Calc. Vol. _____

Meas. Vol. _____

Core Vol. _____

Drive Shoe Factor _.66_

Core Factor _0.52_

Vol. Factor _3.1_

Est. Wt. Mgs. * _____

Wt. Gold. Mgs. _324.1_

Correction _24.1 x 3.1 186.3 = 123_

Corrected Gold. Mgs. _2815_

Est. Fineness _800_

U. S. $ per Fine Oz _35.00_

Wt. Black Sand Oz _____

Bedrock elev. _45.85_

Normal W. L. _70.1_

Dredging Depth

Below Normal W. L. _25.6_

PERSONNEL

Driller * _Eng. Office_

Foreman * _____

Calc. by _____

Checked by _____

Engineer in Charge _____

* These entries must be completed in the Field.

See over for Remarks

B. R. Class No.

FIGURE 4-DRILL LOG 2

FORMA 94 - ENERO 73 500

Tip. Ideal

Appendix D
BL/M Dredge Flow Diagrams/Processing Plants

Figure 1-Flow Diagram/Processing Plant, 20 ft³
BL/M Dredge, W/Cleaveland Circular Jigs

Figure 2-Flow Diagrams/Processing Plant, 14 ft³
BL/M Dredge, W/Cleaveland Circular Jigs

Figure 3-Portable Wash Plant 165 m³/hr,
Cleaveland Circular Jigs

Figure 4-Flow Diagram/Processing Plant, 18 ft³
BL/M Dredge, W/Yuba Jigs

Figure 5-Flow Diagram/Processing Plant, Yuba
18 ft³ Dredge, W/Cleaveland Circular Jigs

Figure 6-Flow Diagram/Processing Plant, 20 ft³
BL/M Dredge, W/Cleaveland Circular Jigs

Figure 7-Flow Diagram/Processing Plant, 14 ft³
BL/M Dredge, W/Cleaveland Circular Jigs

Figure 8-Flow Diagram/Processing Plant, 14 ft³
BL/M Dredge, W/Cleaveland Circular Jigs

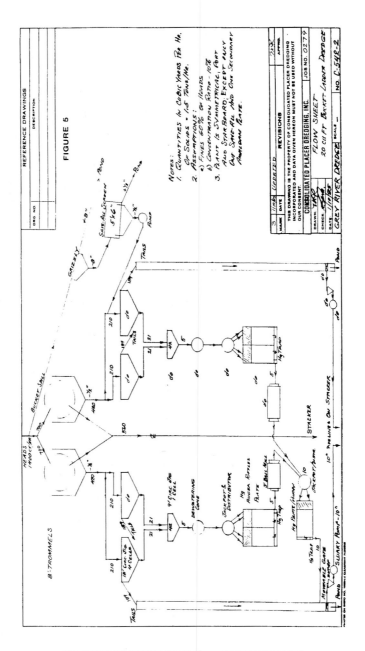

**FIGURE 1-FLOW DIAGRAM/PROCESSING PLANT,
20 FT³ BL/M DREDGE, W/CLEAVELAND CIRCULAR JIGS**

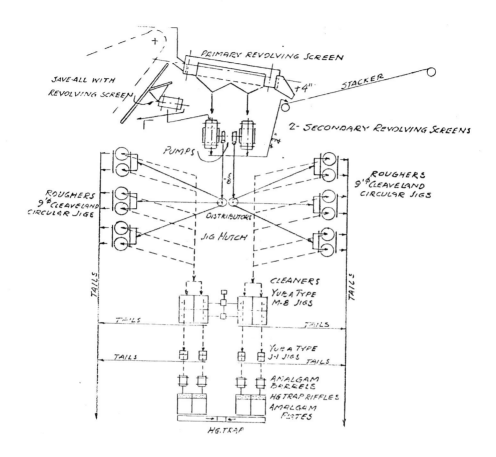

FIGURE 2-FLOW DIAGRAMS/PROCESSING PLANT,
14 FT³ BL/M DREDGE, W/CLEAVELAND CIRCULAR JIGS

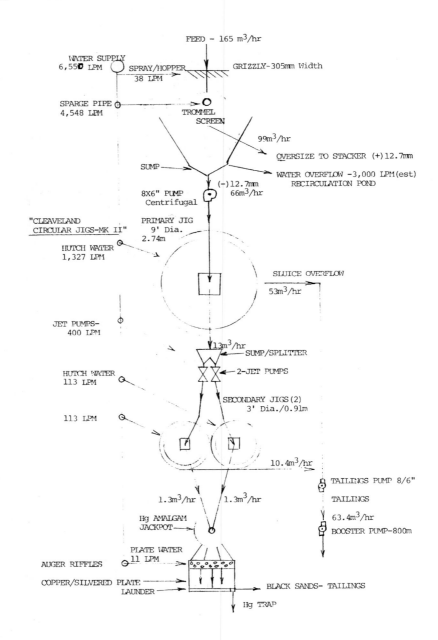

FEED – 165 m³/hr

WATER SUPPLY
6,550 LPM   SPRAY/HOPPER   GRIZZLY-305mm Width
             38 LPM

SPARGE PIPE   TROMMEL
4,548 LPM   SCREEN

99m³/hr

OVERSIZE TO STACKER (+)12.7mm

SUMP

WATER OVERFLOW –3,000 LPM(est)
RECIRCULATION POND

8X6" PUMP   (-)12.7mm
Centrifugal 66m³/hr

"CLEAVELAND
CIRCULAR JIGS-MK II"   PRIMARY JIG
                        9' Dia.
                        2.74m
HUTCH WATER
1,327 LPM

SLUICE OVERFLOW

53m³/hr

JET PUMPS-
400 LPM

13m³/hr
SUMP/SPLITTER

HUTCH WATER   2-JET PUMPS
113 LPM

SECONDARY JIGS(2)
3' Dia./0.91m

113 LPM

10.4m³/hr

TAILINGS PUMP 8/6"

1.3m³/hr   1.3m³/hr   TAILINGS

63.4m³/hr

Hg AMALGAM   BOOSTER PUMP-800m
JACKPOT

PLATE WATER
11 LPM
AUGER RIFFLES

COPPER/SILVERED PLATE   BLACK SANDS- TAILINGS
LAUNDER

Hg TRAP

FIGURE 3-PORTABLE WASH PLANT 165M³/HR-CLEAVELAND CIRCULAR JIGS

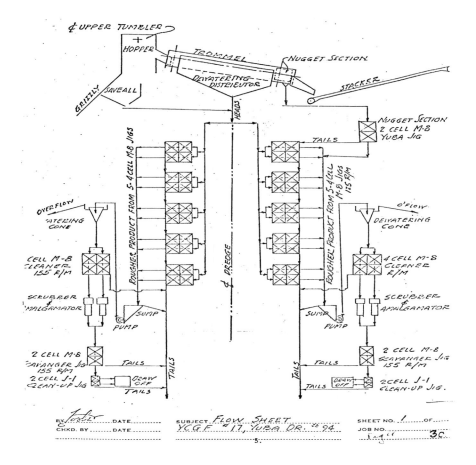

FIGURE 4-FLOW DIAGRAM/PROCESSING PLANT, 18 FT³ BL/M DREDGE, W/YUBA JIGS

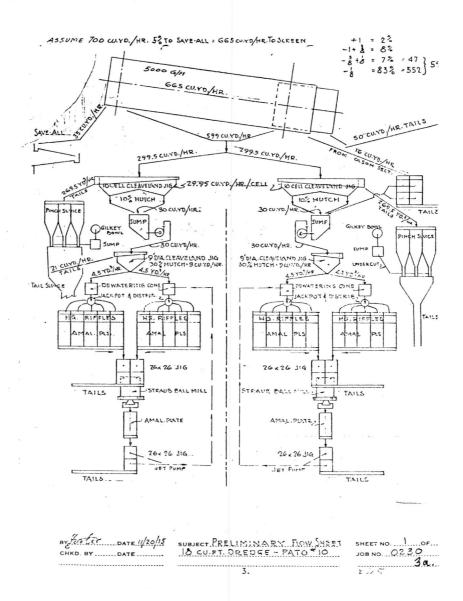

**FIGURE 5-FLOW DIAGRAM/PROCESSING PLANT,
YUBA 18 FT³ DREDGE, W/CLEAVELAND CIRCULAR JIGS**

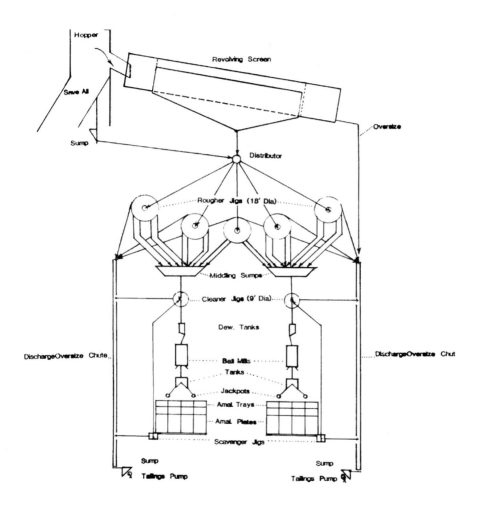

FIGURE 6-FLOW DIAGRAM/PROCESSING PLANT,
20 FT³ BL/M DREDGE, W/CLEAVELAND CIRCULAR JIGS

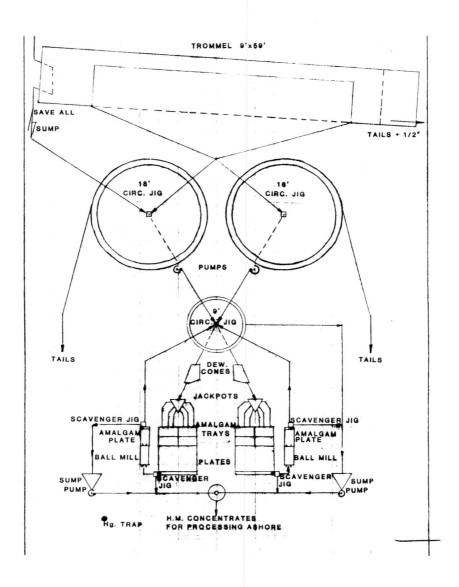

***FIGURE 7-FLOW DIAGRAM/PROCESSING PLANT,
14 FT³ BL/M DREDGE, W/CLEAVELAND CIRCULAR JIGS***

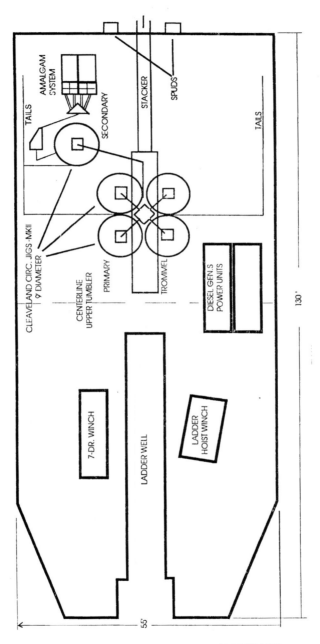

FIGURE 8-FLOW DIAGRAM/PROCESSING PLANT,
14 FT³ BL/M DREDGE, W/CLEAVELAND CIRCULAR JIGS

CASE STUDIES

Appendix E

HISTORY OF MINING ALLUVIAL GOLD
by
Charles M. Romanowitz

Biblical References

The mining of alluvial gold was the first metal mined, recorded in the Bible in Genesis, Chapter II, Verses 10-12 inclusive.

These read;

(10) "And a river went out of Eden to water the garden and from thence it was parted, and became into four heads.

(11) The name of the first is Pison: that is it which compasseth the whole land of Havilah where there is gold.

(12) And the gold of that land is good; there is Bdellium and the onyx stone."

Bdellium being either a precious stone or gum of some tree along with onyx seems to clearly identify this gold to be alluvial.

An outstanding paper by Leon Dominian of New York, entitled "History and Geology of Ancient Gold-Fields in Turkey," (pages 569 through 589, AIME Vol. XLII for 1911), gives an interesting story on the history of gold including reference (page 586) to the above mentioned biblical record. This paper gives the probable location of the Pison River in Asia Minor in the Pontic Gold Fields, one of three areas he discusses to a great extent.

Also of interest Dominian found in his research that "At the height of the Macedonia Power, shortly after 400 B.C., there was the prevailing belief that if gold was extracted by the pick it would immediately grow again like grass mowed by the Scythe." Too bad that was only a myth, for otherwise we would not need the exploration cycle in placer mining. It did, however, pinpoint the use of the pick but neglected the associated tools, the shovel or whatever else they found necessary to assist in obtaining the gold. Since early prehistoric times, gold has been in great demand and sought all over the globe by mankind, first probably as ornaments, then for regular services of all religions, then for monetary, industrial, arts, sciences and, not to be neglected, hoarding purposes; gold being long-lasting along with its malleability apparently established its desirability for many purposes.

Of the alluvials, gold must be considered first herein for it led to the use of the dredge, the subject of this treatise.

Native Hand Mining

Natives' desire for gold led to the developing of equipment to mine for and obtain gold. The first equipment after the pick and shovel used world-wide for this purpose was the wooden batea, or pan, being of many shapes and sizes. No records are available of the evolution that led to the batea, but of interest is the fact that these were developed singularly and very similarly by aggressivenatives around the world in order to increase their individual wealth and, without a doubt, to gain favor with their women due to the love for gold usages. With the women's power to influence man it must have played a great part in the development of early placer mining.

Metal Pan Introduction

From the wooden batea evolved the metal pan with shape and proportional dimensions copied in many cases from the original wooden models. The metal pan afforded an opportunity to add a copper lining, permitting quicksilvering the pan for more efficiently obtaining gold by amalgamation.

The most important fact is that the use of panning from the start was a gravity concentration, and the principle is the same today for all alluvial treating systems. By the very nature of the alluvial formations and the ores contained therein, no other treating system has been possible.

The use of pans alone was of little value without means to hand mine, making possible the selection of the desired ore-bearing gravels. The pick and shovel in their various forms were the tools used for centuries. Then the pans, and from the pan evolved the sluice and rocker for greater production and lower costs, with the ore-bearing gravels excavated by the use of pick and shovel.

Wheelbarrows for Alluvials

Greater production was demanded and was next accomplished by increased treating surfaces in progressive steps as both yardage handled and sluice capacity increased, being supplied with alluvials first by shovel-filled wheelbarrows then the horse-drawn scrapers or similar tools. This latter was the first advance from hand excavation operations which then evolved into powered equipment for excavating for what was then considered mass production.

Hydraulic Monitors

On some placers where conditions and environments permitted, especially with suitable elevations and high water pressure the excavation and transportation of ore-bearing gravels was accomplished by the use of hydraulic monitors.

This developed into the flourishing hydraulic mining industry. It moved great areas of hillside mining formations and produced very large production of gold at extremely low costs. However, it was stopped generally by legislation in the 1880's or reserve areas being worked out.

California Placer Mining in 1800's

The simple hand mining operations in California alone flourished for many years. The yearly gold production in 1852 proved to be the greatest in California state's history regardless of what methods were used thereafter.

In that year 3,932,631 ozs. of gold were produced with a value of $81,294,700. when gold was valued at $20.67 oz. Even with the large capacities produced by dredges of all types, the maximum yearly gold production in California in 1940 after gold was revalued was 1,455,671 oz's valued at $50,948,585. A sizable reduction from the hand mining.

Mining Dredge Introduced

The mechanical mining dredge did not make its successful appearance for over three centuries after the first "Bag and Spoon" type dredge, used only for excavation, made its appearance in the lowlands of Europe apparently in what is now Holland.

A drawing dated 1565, is shown on Page one of "Dredging and Dredge Appliances," by P. M. Dekker, published by The Technical Press Ltd. of London in 1927. The Grab Bucket dredge is shown on Page 2, dated 1617. From these illustrations it is evident that the original dredge for underwater excavation was in the early part of the 16th Century.

Dr. M. Fritzlin in his article, "Development of the Dredging Industry," pages 15-19 inclusive, World Dredging and Marine Construction magazine, Vol. 2, No. 4 Nov/Dec. 1966, gives an excellent account of the development of the dredges as excavating units. He dates the start of excavating units centuries before Dekker does.

In the interim between the "Bag and Spoon" dredge to the first successful Bucket Line mining dredge, gradual progress was made in improving the efficiency of the dredge as an excavator, resulting in the continuous-bucket, Bucket Line type, the Grab Bucket, as well as similar types, and finally the Suction or Hydraulic type being the last to appear, in 1864 (see "Hydraulic Dredges," by J. E. Yager in the Excavating Engineer, Page 15, March 1951, Bucyrus-Erie Co., South Milwaukee, Wisconsin, publisher).

The greatest development of dredges for excavation, according to records, occurred in the Lowlands of Europe due to the environment dictating the necessity of using a mechanical unit to excavate for waterways and at the same time to reclaim land for various uses, as well as to protect those lands already reclaimed by making dykes or levees around them.

Alluvial mining by hand thrived for many centuries, first starting in Asia Minor then later in other areas such as in Spain, (see "Gold Dredging," by T. C. Earl, 1913, published by E&FN SPON, Ltd., 57 Haymarket, London S.W.), the alluvial hand mining by the Romans and Moors, also in Hungary, China, Siberia, Australia, and many other countries on record indicated large volumes having been mined. It remained however for the miners in New Zealand to start the mechanization of the alluvials with floating equipment.

From the Lowlands of Europe the art of reclamation, making of new lands in watery areas, as well as waterways or deepening same for water transport and for drainage purposes spread to other areas in Europe and eventually all over the world. Just as civilization started in the Middle East, then to Europe, spreading both eastward and westward to finally meet in California, so did the art of dredging for other than mining purposes move, arriving eventually in California. Placers Discovered in Australia & New Zealand

After alluvial gold was discovered in Australia and California, then later in New Zealand, it was the New Zealanders, including some experienced hand mining Californians who went there expecting greater riches, who made the first mechanical use on record for the mining of alluvial gold.

Gold was discovered in Australia in 1823, in California prior to the famous John Marshall discovery of 1848. Gold was not discovered in New Zealand until 1852, by Charles Ring, a Californian, in the Coromandel Gold Fields of the North Island.

(see Page 52, "The New Zealand Mining HandBook," published in 1906 by the Mining Bureau editor, P. Galvin and printed by John Mackay).

The Charles Ring strike in New Zealand, did not flourish due to hostile natives and severe environmental conditions. Similar to the strikes prior to 1848 in California, the New Zealand hand production had to await the discovery in 1857 in the Collingwod district, followed by several discoveries in other fields up to 1865, to attain success (see handbook reference above, Pages 7 to 17 inclusive).

Some of these New Zealand finds were made by Californians, in fact the most productive by Hartley and Riley (Reilly) on the Clutha River in 1862. They were rewarded years later by having their several dredges operating in the same location, which was unique as the first such case on record of a personaltie-in between hand and mechanical operations.

After the first effective gold find in California-in 1848, which reached its peak in 1852 when the first discovery was made in New Zealand, it was those miners from California who migrated globewise north, east and westward, who apparently were the nucleus for the hand operations elsewhere. Some of these miners also were the aggressors as far as mechanization was concerned, as in the case of Hartley and Riley.

Why it took so long for the miners in the USA, and especially California, to mechanize the operations is a moot question. In California on the Yuba River in 1858, the Phoenix, a steam-driven river boat, tried unsuccessfully to mine gold placers and, as far as theauthor knows, no other attempts were made until early in the 1890's. This may have been due to many of the placers being not as water bound as in New Zealand.

The New Zealanders first adapted the European "Bag and Spoon" typedredge, readapted to make their first Spoon-type dredge (no Bag) in the early 1860's. It was usuable only in shallow and quiet waters, but a number were used.

Next was developed the Current-Wheel dredge in 1868, (see Page 244 of the 1906 New Zealand Mining Handbook mentioned above) capable of limited production in fast moving current streams, but was ineffective along the river beaches or areas in the water where the current was slow, which led to the steam-powered dredge in 1881 (see Page 245 of the above handbook).

A large number of the current-wheel dredges were used and being made of two long pontoons separated for a continuous wellhole to permit the use of the ladder. One current-wheel was placed on the outside of each pontoon to supply the power to drive the bucket line only, with hand driven winches to raise and lower the bucket line ladder and operate the mooring lines. All dredges used headlines.

These operations being limited led to the use of the steam powered dredges. In 1887 the first suction-type dredge was put into service near Alexander, New Zealand, followed by several others, but these operations were limited to fine sand formations and in time were replaced by the bucket-line dredge. This first operation of the suction-type dredge was carried on by a Chinese merchant. This type dredge was the first resoiling-type dredge used, but was limited to certain formation conditions, especially to only a few feet of bank ahead of the dredge.

In 1890 the first application of electric power for dredges took place. In 1894 the first Stacker, the Cutten Elevator, was used to replace theflumes or chutes to dispose of the oversize gravels from the end of the screen. The Stacker, using light steel buckets, was very short and steep. All the above dredges, especially the suction-type, experienced great difficulties capturing the fine gold (all the above data from Pages 241-319, of the 1906 New Zealand Handbook).

By 1900, there was reported over 200 dredges operating "down under," or being built for operation in New Zealand and other countries such as Australia, Canada and the U.S.. The designs and manufacturing rights for the New Zealand-type dredge for the U.S.A. were obtained by the Risdon Iron Works of San Francisco, a forerunner of one of the divisions of the Bethlehem Shipbuilding Company.

While all the above progress was being made in New Zealand, it was not until 1895 that the first bucket line mining dredge was placed in successful operation in the U.S., in Montana. It was the design used for excavation in the eastern U.S. and adapted for mining purposes by the addition of the washing and treating systems.

In California a New Zealand-type dredge had been tried before 1898, but was unsuccessful, reportedly due to lack of gold values. The same type dredge was installed on the Feather River near Oroville, California, to start the first successful mining operation in January 1898. This was done by W. P. Hammon with the financial assistance of Tom Couch, who came from Montana, and before any finances would be forthcoming Couch insisted on certain amounts of prospecting to insure success, which was not done in the case of the earlier attempt.

Soon thereafter, the type of bucket line dredge used in Montana appeared in California and was the basis of the dredge developed as the California-type, the most successful mining dredge used, first in California then in various alluvial areas around the world. By 1900, there were many in operation in California and shortly thereafter, the dredging boom was spread all over the west and north into Alaska and western Canada.

Because of the great variety and widest range of formations and environmental conditions, including weather, the greatest development of the dredge took place in California. There were several manufacturers in the USA very prominent in the design and manufacture of the bucket line dredge. The first was the Bucyrus Co. (later became Bucyrus-Erie), represented in the west by the Yuba Construction Co. Link Belt Co. and Marion Steam Shovel Co. also supplied many of the early dredges.

After 1909, Yuba started to design and produce their own dredges and finally cancelled out the Bucyrus representation in 1913. The globe. Most or these dredges used spuds, instead of headlines, to more firmly anchor the dredge. The spud-type dredge could also operate in all heights of banks beyond that possible for headlines, as well as producing the largest unit yardages up to about 1965.

Yuba was also the most called upon to furnish personnel for dredges of many makes and designs due to direct contact with varied operations until the late 1950's.

The New Zealand Headline type dredge, with some adaptations of certain units from the California-type design, was used in many fields where it was found to be more practical. Applications in the Far East included Malaysia, Thailand, Burma and Indonesia. These dredges were generally designed by the F. W. Payne & Sons, or the several Holland firms now making up the IHC-Holland organization, the latter being the largest manufacturers at this time (1975).

Manufacturing was mostly done in Great Britain. Since WWII, the above companies were active, such as Simon Lobnitz Co., and have produced the most sophisticated and efficient dredge of today, embodying modern designs, producing new record high yardages and low costs.

The suction or hydraulic mining dredge, as mentioned above, was first used for mining purposes in New Zealand. Due to its limited scope of action and inefficiency as a fine gold saver, it was replaced by the bucket line dredge. In the USA, the suction (hydraulic) dredges were also tried for mining without success due to the same general reasons as in New Zealand. During the 1930's several suction dredges were used in California and other western states, but little was given out as to their operating cost and efficiency.

Doodle-Bug Placer Mining System
With the depression of the 1930's, along with the revaluation of gold(to $35/Troy oz), the non-mining excavating industry found itself in a dilemma, with excellent equipment and experienced operators idle, so turned to placer gold mining. This was mostly done by draglines which were put to work excavating the ore-bearing uncon-solidated formations, but had to use separate units alongside for the washing and treating systems.

These separate units adjacent to the dragline were generally floating but in some cases the environment and for other reasons, dictated the use of them being non-floating, using caterpillar track units for portability. These combinations were called "Doodle-Bugs" and played an important part in the gold production, up to the start of World War II.

The revaluation of gold produced a great mining boom in every gold bearing area. The Doodle-Bug operators aggressively vied with the bucket-line industry to secure mining rights in areas to mine placer gold and, later, other minerals. Areas never known before to have profitable values were discovered and developed and in some cases the Doodle-Bug operators found areas not considered placer mining possibilities bv the bucket line operators, as some having very fine values.

Unfortunately, due to severe competition even among the Doodle-Bug industry, many areas were cut up into too small parcels that hurt the overall possibilities of profitable mining. In some cases the bucketline operators, if controlling some of these rights in adjacent areas, could have made a much more profitable project. It was at this time that the portable bucket line dredge was first developed by Yuba and produced by others, making possible the profitable mining of smaller areas than the usual conventional bucket lines.

The Doodle-Bug industry produced a large amount of placer gold and some other metals but were not able to operate as efficiently or at as low a cost as the bucketline dredge. The early 1940's saw this new 1930 industry disappear as an important mining industry. In many areas the scope of action of the Doodle-Bug operated profitably under environments not possible with even the portable bucket line units.

These included conditions such as shallow, narrow, steep, bouldery in proportion to the depth; slope and small yardages; all if values were in range with the difficulties involved. During the peak of its operations in the 1930's it was reported that as many as 150 units operated on the west coast.

The Bodinson Mfg. Co. of So. San Francisco, was the leading designer and manufacturer of Doodle-Bugs and served the industry well. An unique Doodle-Bug operation was the one on a placer in the James Creek, Napa County, California. This operation was for the mining of cinnabar, using a Bodinson washing plant (see Mining Journal, Phoenix, Arizona, August 30, 1945 Issue, "Test Run of James Creek Cinnabar Dredge," by P. Patton).

During the same period(1930's), of the high activities of the Doodle-Bug, a few Suction dredges were put into operation. Due to the excessive quantities of water inherent with this type of dredge, its efficiency as a placer mining unit was low, produced particularly by the dewatering problem and its scope of action limiting its use.

Today that situation is far different due to the use of the new Cleaveland Circular Jig. The suction dredge has fast become a competitive mining tool, especially offshore. If the Union Carbide's subsidiary TEMCO, to be employed offshore of Thailand, proves up to prediction this type of dredge will be used in great numbers in the future.

After the start of the 1940's, and especially when WWII started, the governmental restraining order L208 was put into effect. It unjustly discriminated against the gold mining industry as it did not serve its purpose of forcing the gold miners into wartime strategic metal mining industry. The industry never recovered from its bad effects.

Subsequently, the higher prices of labor and materials prevented the restart of many of the gold placer mining operations. This was a practical demonstration that gold mining is at least a semi-depression industry thriving best when other industries are adversely affected in depression periods.

To date the placer gold mining has still not come back and all dredging for such has practically become extinct in the USA and Canada, except for a few isolated cases in Alaska. The one placer platinum group and with gold as a by-product bucket line dredging operation, has continued uninterrupted at Platinum, Alaska, by the Goodnews Bay Mining Co., the largest producers of the alluvial platinum group in the USA. This operation has been a continuous one since its start in late 1937.

Operations have continued in South America for both placer gold and the platinum group, in Colombia, Bolivia, Peru and Brazil. The largest operations for these metals are those of the International Mining Corp. of New York with eight dredges in Colombia and one in Bolivia.

During WWII, the bucket line tin ore (cassiterite) dredging industry all over the globe was badly crippled and retarded. Since the end of WWII, while this tin mining industry has had period of good and bad times, generally it has been good. Enjoying the highest prices in history, tin has enabled many areas that were mined before, when prices were low and equipment with low efficiency, to be re-mined.

In some cases operating in areas of previously two time operations with the present day using generally highly improved equipment including more efficient treating systems permitting formations with tin content as low as 0.2 lb/cuyd, which is far below that heing possible even as late as the 1940's. This equipment has all been bucket lines as well as all offshore in Indonesia and Thailand. The suction dredges have been used onshore in Indonesia since before WWII (for stripping of overburden).

Background of YUBA CONSOLIDATED GOLDFIELDS CORP. at HAMMONTON, California; Leader in Placer Gold Dredging
This operation was the largest alluvial gold operation of any single field in the free world. Along with it there were also three smaller operations in the same adjoining areas.

Before the hydraulicking of the alluvial gravels were stopped in the 1880's, the Yuba River carried during floods a large quantity of tailings and deposited them on portions of the YCGF area, which in some places totaled about 30 feet. These tailings carried some fine gold and mercury from the hydraulicking sluices.

The deepest depth of gravels extended to approximately 200 feet below the surface to bedrock. As the depths of gravel were greater than the digging depths of available dredge designs, some of the gravels were redug as much as three times, as the dredges were able to go to greater depths (in successive years of development).

The YCGF over a life of 64 years (1904-68), dug up 1,081,000,000 yd³ of gold bearing gravels, having dredged recovery value of $0.1275/yd³, at a cost of approximately $0.063/yd³ (field operating costs), and a gross return of almost $138 million. By the end of 1938, the total dividends were over $30 million. No record is available of the additional dividends between 1937 and October 1, 1968, when the last dredge, Yuba Gold Field Co.#21, was shut down.

The other dredging companies were:

1. The Marysville Gold Dredging Co., which was finally changed to Marigold Drilling Co.. They sold their last dredges, #5, to YCGF who renamed it YCGF #19, an 18 ft³ dredge digging to 81 feet below water. Its property was on the lower- or western-end of the YCGF's, but just adjoining the river on the south side. It is estimated that this property contained at least 300 million cuyd's.

2. The Yukon Gold Co., now Pacific Tin Consolidated Corp., had one dredge, 9 ft³ capacity, on the river near the upper end.

3. The Williams Bar Gold Digging Co. had one smaller size dredge, of about 6 ft³ on the extreme upper end of the YCGF property.

All told, it is estimated that all the dredging operations total about 1.400 billion yd³ that were dredged.

The YCGF Co. had a total of 22 dredges, 4 of which were rebuilt by Yuba Manufacturing Co., and equipped with steel hulls. The first dredges dug to a depth of approximately 60 ft. Then when Yuba #16, 17, and 18 were built, they dug to 81 ft, as did Yuba #19, from Marigold. Then Yuba #17 was changed to dig to 112 ft; Yuba #20 to 124 ft; and Yuba #22 to 107 ft. Yuba #22, the last, was the smallest capacity and digging depth, 6 ft^3 buckets and 26 ft digging depth, both because the area was shallow and small in total yardage.

It is of interest that this dredging field has been the training grounds for dredge operators sent to dredging areas worldwide. Also, many of their dredges have been shipped to their other dredging areas in the US, and others have been sent to dredging areas worldwide. YCGF had many subsidiaries located in many areas in the US, including Alaska and one in Portugal.

End.

BIBLIOGRAPHY

Agricola, Georgius, "De Re Metallica," c. 1556, translated by
Herbert C. & Lou Henry Hoover, 1912, re-published 1950,
Dover Publications, Inc., New York, NY.

Aubury, Lewis E., 1905," Gold Dredging in California,"
The California State Mining Bureau, Sacramento.

Averill, Charles V., 1946, "Placer Mining for Gold in California,"
Bulletin 135, State of California, Dept of Natural Resources,
Division of Mines.

Barber, Lee, "Establishing Dredge Mining Operations in Certain
Isolated Locations," Ocean Mining Symposium(OMS),
M.J.Richardson Inc., PVE, Calif.

Becker, W.R., 1976, " A Modern Approach to Placer Exploration
Drilling & Sampling," Becker Drills Co..

Breeding, W., 1978, "Tin Placer Mining Methods," May, WDMC.

Carlson, C.G., 1978, "Exploration and Evaluation of Placer Mineral
Deposits," Dredging Technology Seminar, University of Nevada,
Reno, NV.

Carpenter, J.H., 1952, "Mining & Concentration of Ilmenite and
Associated Minerals at Trail Ridge, Florida," SME/AIME.

Carr, J.
1972, "Some Aspects of Gold Dredging in New Zealand-
In Retrospect," WODCON.
1974, "Dredging for Gold in New Zealand," WODCON.

Clauss, G., 1972, "Airlift Systems for Mineral Recovery in Ocean Mining," WODCON.

Cleaveland, Norman
1967, "Some Comments on Dredging as a Mining Method," WODCON ASSN, PVE, Calif.
1967, "Systems for Ocean Mining," OMS.
1970, "What Tin Mining Has Meant to Malaya," WODCON.
1972, "Dirt, Diamonds and Gold Dust," World Dredging & Marine Construction,(WDMC), San Pedro, Calif.
1973, "Bucket Dredge Systems for Mining-Design Innovations," Ocean Mining Symposium, WODCON.
1973, "Bang! Bang! in Ampang; Dredging Tin During Malaya's Communist Emergency," Symcon Publishing Co., Irvine, Calif.
1976, "Some Proposals to Improve Placer Dredging," WODCON.
1978, "Dredge Mining-More Tin at Lower Cost," WDMC.
1979, "Let's Stop Wasting Tin," WDMC.
1980, "New Zealand to the Rescue," WODCON.
1982, "Circular Jig," U.S. Patent No. 4,310,413.

Cottell, D.S., 1981, "Offshore Recovery of Sand & Gravel-An Alternative to Inland Mining," Nov., WDMC.

Daily, Arthur, 1962, "Valuation of Large Gold Bearing Placers," Engineering & Mining Journal, Vol. 163.

Dieperink, J.H.J., 1977, "Large Dredge Designed to Mine Indonesian Tin," WDMC.

Dooremalen, J. van, 1974, "New Developments in Integrated Processing Systems on Sand & Gravel Dredgers," WODCON.

Donkers, J.
1967, "Equipment State of the Art, European Dredging Industry" OMS.
1970, "Some Aspects of Modern Alluvial Mining Equipment," WODCON.

Dunkin, H.H., 1950, "Operations of Bulolo Gold Dredging Ltd.," University of Melbourne.

Earl, T.C., 1913, "Gold Dredging," E&FN Spon, Ltd., London.

Gardner, Charles W., 1921, "Drilling Results and Dredging Results I & II," Vol 112, Engineering & Mining Journal.

Goodier, J., 1967, "Dredging Systems for Deep Ocean Mining," WODCON.

Harding, James E., 1952, "A New Way to Interpret the Results of Placer Sampling," Engineering & Mining Journal.

Harrison, H.L.H., 1962, "Valuation of Alluvial Deposits; Dredging for Gold and Platinum-Alluvial Mining for Tin and Gold," Mining Publications, Ltd., London, UK.

Healy, A.M., 1967, "A History of the Development of the Bulolo Region, New Guinea."

Herbich, John B., 1975, "Methods for Offshore Dredging," WODCON.

Hidalgo, I., 1970, "Mini-Dredge (Iron Sand Mining) in the Philippines," WODCON.

Hill, J.C.C., 1976, "Economic Evaluation of a Marine Alluvial Deposit Using Vibratory Coring," MacKay School of Mines, Reno, NV.

Hutton, Gerald H., 1921, "Placer Prospecting Practise," No. 176, Mining & Metallurgy.

Jennings, Hennen, 1916, "The History and Development of Gold Dredging in Montana," & Chapter on "Placer-Mining Methods and Operating Costs" by Charles Janin, Washington GPO.

Kastelic, W., 1978, "Ground Preparation for Alaskan Gold Mining Ventures," May, WDMC.

Leaver, E.S. & Woolf, J.A., 1937, "Gold Dredging in California and Methods Devised to Increase Recovery."

Malozemoff, P., "Jigging Applied to Placer Gold Recovery," 1937, Engineering & Mining Journal.

McDowell, A.W.K., 1980, "The Underwater Bucketwheel Excavator-A Review of 15 Years," May, WDMC.

McGeorge, J.W.H., 1964, "Dredging for Gold," Brown, Prior, Anderson Pty Ltd, Melbourne.

Misk, A., 1976, "Diamonds Lend Sparkle to Dredging Operations," WODCON.

Miyake, Atsusato; Ofuji, Ichiro et al, 1989, "Development of Automatic Operation System Incorporating Fuzzy Control for Cutter Suction Dredge," WODCON.

Neil, J.W., "Application of Jigs to Gold Dredging," 1914, Mining & Scientific Press.

O'Neill, Patrick H.
c. 1964, "Gold and Platinum Dredging in Colombia and Bolivia, South America," International Mining Corp., New York, NY.
1977, "Placer Mining With Bucket Ladder Dredges," WDMC.

Richardson, M.J.
1980, "Dredge Mining of Placer Gold in the 1980's," WODCON.
1982, "South American Dredging Mining," July, WDMC.
1983, "Handbook of Mineral Jigs," Consolidated Placer Dredging, Inc. (CPD).
1984, "Handbook of Alluvial Gold Evaluation Methods," (CPD).
1984, "The Evolution and Current Applications of the Mk.II Cleaveland Circular Jig to Alluvial Gold Recovery," July WDMC.
1985, "Placer Gold Mining: A Return to Basics," July, WDMC.
1986, "Evaluation/Decision Process for Small-Scale Placer Gold Mining," April, MINING MAGAZINE, London, England.

Richards, Robert H., 1908-9, "Ore Dressing" (4 Volumes), McGraw Hill Book Co., New York, NY.

Romanowitz, C.M., Circa 1966, "Onshore Alluvial Mining Results As A Guide to Future Offshore Mining," US Bureau of Mines, San Francisco, CA.

Shlemon, Roy J., c.1970, "Geomorphology of the Lower Nechi Valley," Pato Cons. Gold Dredging Ltd.

Smith, Howard D., 1916, "Gold-Saving on Dredges," Mining & Scientific Press.

Smith, R.G., 1921, "The Discrepancy Between Drilling and Dredging Results," Vol 112, Engineering & Mining Journal.

Stanaway, Kerry J., 1990, "Heavy Mineral Placers," BHP-UTAH INTERNATIONAL INC., Herndon, VA.

Taggart, Arthur F., "Handbook of Mineral Dressing-Ores and Industrial Minerals," 1945 Edition, pg's 2-90/91, John Wiley & Sons, New York; Chapman & Hall, London.

Vedensky, D.N., "The Application of Jigs in Placer Mining Operations," 1938, Pan-American Engineering Co., Berkeley, Calif.

Weatherbe, D'Arcy, 1907, "Dredging for Gold in California," Mining & Scientific Press, San Francisco.

Wells, John H., 1969, "Placer Examination, Principles and Practice," U.S. Dept of Interior.

Wheeler, B., 1980, "ARC Marine Expanding in Lucrative UK Offshore Aggregate Market," WDMC.

Wilson, Eugene B., 1898, "Hydraulic & Placer Mining," John Wiley & Sons, New York; Chapman & Hall, London.

Wimmler, Norman L., 1927, "Placer-Mining Methods and Costs in Alaska," US GPO, Washington.

Wolff, Ernest, 1964, "Alaskan Prospector, Handbook for the," University of Alaska.

Yevgeny I. Bogdanov, 1990, "Placer Mining Technologies in the USSR," Alaska Miners Assn.

INDEX

Aerial, Surveys 3
Alaska 4, 15, 27-8, 49, 52, 143, 189, 226
Alaska Gold Co. 190
Alcon Ltd. 39
Alluvial deposits 42,49
Amalgamation (see Mercury)
American Bureau of Shipping (ABS) 149
ASARCO 52
Australia 143, 229
Automation 172-3

Bade Drill (see Drill)
Ball Mill 188
Banka Drill (see Drill)
Barite 183
Batea 5, 29
Becker Drill (see Drill)
Bedrock 1, 2
Bolivia 25, 228
Brazil 25, 143, 193-4, 226
Bucyrus-Erie 143
Bulolo 1, 189, 197, 227
Bureau Veritas 149

Caisson Drill (see Drill)
California 1, 12, 18, 25, 50, 143, 229
Canada 4, 15, 50, 143, 189
Cassiterite (see Tin)
China 4
Cleaveland, Norman 192-6, 205
Coal 183
Coco matting 23

Colombia 143, 225
Color estimation 29 (see Log)
Computer
    Program 38
COMSUR/SAPI 228
Copper 183
Crangle 182 (see Jig)

Dead Sea, Jordan 230
DeBeers Laboratory 193
Desert placers 51
Diamond, placers 3, 40, 221, 226
Distributor (see Dredge, Subsystems)
Dredge, CATEGORIES 142
    Bucket LaddeR (BL/M) 4, 88-9, 91-3, 97-8, 141-57, 173, 187, 193,
                                    225-30
    California-type 143-4, 151
    European-type 144, 151
    Bucket Backhoe (BB/M) 24, 166
    Bucket Clamshell/Grab (BC/G/M) 165
    Bucket Dragline (BD/M) 165, 167
    Bucketwheel Suction (BWS/M) 92-3, 168-70
    Cutter Suction (CS/M) 92, 170, 229
    Plain Suction (S/M) 92, 171, 229
    Trailing Hopper Suction (TH/M) 92, 171, 229
    Dustpan Suction (DS/M) 230
Dredge, OPERATIONS 225-30
    Bucketline Speed 146, 148
    Mining Course 221-2
    Production 146-8, 168
Dredge, Size 147
Dredge, SUBSYSTEMS 147
    Buckets 150
    Digging System 150-1
    Distributor 157
    Electronics 172
    Hopper 153
    Hull 148-9

Jackpot System (Hg) 159-62 (see Mercury)
Jig (see Jig)
Save-All 156-7
Trommel, Revolving Screen 7, 94, 153-5
Drill, CATEGORIES
  Auger 10
  Banka 8-11, 43, 45
  Becker, Hammer RC 8, 15, 27, 43, 47, 49, 52-9
  Buycyrus-Erie 13-4
  Caisson, Bade 25
                Yost 26
  Churn 9, 13-4, 27, 52
  Empire 9, 10
  Engine-powered 12
  Hillman, "Airplane" 13-4
  Keystone 8, 12-3, 45, 47, 52-9
  Koehring 14
  Loomis 14
  Rotary 8, 18
  Vibratory 8, 18-9, 28
  Ward 11, 14
  Yost Klam 26
Drill, components 10-2
Drill, operation 9-18
Drilling, Development 44, 46-7
  scouting 42
Drive Shoe 9

Eluvial Deposit 51
English Channel 229
European-type dredge (see Dredge, Categories)

Evaluation
  calculations 88-9, 96, 220 (see Log)
  color estimation (see Log)
  computer 38
  concentration 29
  decisions 4, 90-1, 95-6

    equipment 88-94
    log preparation (see Log)
    procedures 23-4, 87-8
    weighing 29
Exploration 46-7
Fisher & Baumhoff 190
Flow diagram 199
"Fuzzy Logic" 172-3

Geomorphology 42, 246
Ghana 230
Griffin
    Frank 182,188-9
    Maurice 188
Grizzly 93, 155

Hematite 200
Hillman, Kirk/"Airplane" (see Drill)
Holland 229
Hopfield, L.D. 189
Hopper 93
Hughes, A.D. 209
Hydrological/Hydrology 41

IHC Holland 194
Ilmenite 49
Indicated reserves 44-6
Indonesia 143,229
Insurance 149
Jig, Mineral 181
    Applications 187
    diamond 183,193,200
    gold 183, 187-90, 193, 200, 204
    lode 183
    tin 183,188,200
    CPD/PATO Tests 182, 193-4
    Milestones/development 182
    Natomas tests 188

OPERATION 160
bed material 200-2
capacity 160-1, 185, 198-9, 207
cycle 202-3, 205
density of mineral 186, 198
feed
 flow rate 185, 198-9, 202
 ratio of 161
  grain size 186
  hutch, spigot 204
  water 202-3
overloading 185
screen
  analysis 198
  material 201-2
  sizing of holes 186, 201-2
stroke 185, 202-3
tuning 202, 204
Jig,
THEORY
 laws 185
 hindered settling 186
 principles 187, 202
 side-wall effect 185
TYPES
 Basket sieve 182-3
 Baum 184
 Bendelari 155, 189-90
 Bradford, eccentric 184
 Circular, 183, 200, 209
 Bilharz 184
Cleaveland 155-6, 182, 192-6, 206-8, 212-7, 226
Collom 184
Conkling 184
Cornish 184
Crangle 182,189
Diescher 184
Dudley 184

Evans 184
Faust, Henry 184
Ferraris 184
Francon 184
Hancock Vanning 184
Hand operated 184
Harz 155, 182, 184, 191-2
Hodge 184
Hoopers Vanning 184
IHC Holland 194
Luhrig 184
Neill 187-9
Osterspey 184
Pan American 155, 182, 189, 191, 200
Parsons & Fisher 184
Robinson 184
Sheppard 184
Sieve
  Fixed 184
  Moveable 184
Siphon separator 184
Stutz 184
Trapezoidal 195-6
Under-piston 184
Utsch 184
Woodbury 182,187
Yuba 155, 182, 190, 200

Keystone Drill (see Drill)
Klam, Yost Drill (see Drill)
Koehring Drill (see Drill)
Linear Regression 54-5
Lloyd's Register 149
Log, TYPE
  bulk sampling 46
  drill 32-4,36
  shaft 36, 46
  PREPARATION 30

abbreviations 31
color estimations 29-30
calculations 30-5, 37-8, 54-9
corrections 31-7, 55
drive shoe factor 15, 17, 30, 33-5
Radford Factor 36
Long Tom 6, 160
Lyons Syndicate 188 (see Jig)

Magnetite 49,197
Malaysia/Malaya 143, 193, 228
Mercury 23, 29, 30, 95, 158-64
Mineral jig (see Jig)
Mobilization, Dredge 97
Moraine 50

Natomas 40, 50, 143
Neill, James W. (see Jig, Neill)
New Guinea 143,227
New Zealand 15, 50, 52, 143
    Grey River 52-4, 59
    Kanieri Dredge 15
    Taramakau River 15, 52-3, 230
Nigeria 143

Ocean, offshore 49
Orenstein & Koppel 144

Pacific Tin Consolidated Corp. 192, 205, 209
Paleo Channel 41
Pan American jig (see Jig)
Pan, panning 5, 23
Pato Consolidated Gold Dredging Ltd. 40, 197, 225
Payne, F.W. 144, 191, 228-9
Perry, O.B. 188
Peru 50, 227
Phosphate 49
Pitting (see Sampling)

Polygon 220, 231
Processing Plant 25, 88
Proven reserves (see Reserves)

Radford Factor (see Log) 36, 39, 56
Reconnaissance 40-1
Reserves
    Calculation 56, 90, 220
    Indicated 46-7
    Proven 47
Riffles 23, 155, 188, 190, 196-7
Rocker 6, 23, 29
Romanowitz, Charles M. 190, 197
Rotary drill (see Drill)
Russia 143

Salting 21
Sampling
    bulk 19, 24,43-4, 46
    caisson 25-6
    cut box 20-2
    development drilling 44
    pitting/trenching 24
    saturation 39, 46
    scout drilling 42, 44
    shaft 19-21, 23
    spacing holes 24, 44, 46
    surface 42
Save-All (see Dredge Subsystems)
Shot 200
Siberia 4
Simons-Lobnitz 143
Sluice 6
    "Long Tom" 6
Strouse, Edward 189
Survey
    Initial 1-2

Tazmania 143
Tenor (see Log)
Terrace 20, 41
Terrain 28, 41, 43
Tertiary
    deposits 49-50
Thailand 15, 143, 193
Tin 25
Triangular method 220
Trommel, revolving screen (see Dredge, Subsystems)

U.S. Smelting, Mining & Refining Co. (USSM&R) 40, 190, 197

Vibratory drill (see Drill)

Ward drill (see Drill)
Wash plant 25, 88-9, 90, 93
WESTGOLD 49

Yosemite Dredging & Mining Co. 187
Yost Klam Drill (see Drill)
Yuba, Consolidated Gold Fields 40, 190-1, 197
    dredge 143, 194, 228
    river 1, 229
Yukon 189, 229

Zircon 49